Crisis, Reinvention and Resilience in Museums

"With multiple crises unfolding daily around the world, Philip W. Deans provides the reader with an unlikely source of reinvention and resilience—a museum. Deans' work is disciplined, thorough and meticulous, and his message rings with crystal clear clarity—a museum's sustained effectiveness and popularity depends upon its willingness and ability to change. For all organizations, nothing is so simple or profound. If you care about the future of museums, read this book."

—Robert R. Janes, *University of Leicester, UK*

"An excellent and fascinating study on the Imperial War Museum at a crisis point in its history. Built on strong theoretical foundations, this book offers new insights not only into the political, public, and institutional treatment of war in peace and conflict, but also into how museums today can survive crisis. It is well worth reading."

—Emma Cunliffe, *Newcastle University, UK*

"This book gives us a fresh look at museums. Featuring an intensive look at key concepts and important theories, Philip W. Deans helps us rethink cultural institutions, particularly the Imperial War Museum, during an era of great transition stretching from the First World War to the Cold War. This is a study with notable ramifications for museums today."

—Samuel J. Redman, *University of Massachusetts Amherst, USA*

"This meticulously researched study makes a serious contribution to the growing literature on how museums experienced the twentieth century not just as a period of growth, but also as a period of profound instability and threat. Understanding how a museum OF war was impacted BY war adds an important piece to our picture of museum history."

—Kate Hill, *University of Lincoln, UK*

"Despite extensive reading on the history of the Imperial War Museum, and my 32 years of service in the museum itself, this book has been a revelation to me. It is a work that fills a gap in our collective memories of the museum and performs an important service in giving due weight to an era of rapid change and redefinition that does not deserve to be obscured from our view by the shinier redevelopments of the mid-1960s, late-1980s and 2010s."

—Paul Cornish, *Curator (retired), Imperial War Museums, UK*

Philip W. Deans

Crisis, Reinvention and Resilience in Museums

Defence and Revolution at the Imperial War Museum, 1933–1950

palgrave
macmillan

Philip W. Deans
School of Arts and Cultures
Newcastle University
Newcastle upon Tyne, UK

ISBN 978-3-031-67805-9 ISBN 978-3-031-67806-6 (eBook)
https://doi.org/10.1007/978-3-031-67806-6

This Palgrave Macmillan imprint is published by the registered company Springer Nature Switzerland AG.
The registered company address is: Gewerbestrasse 11, 6330 Cham, Switzerland

For Sophie and Eva: 'This thing I did'.

Foreword

In 2022, I had the pleasure of incorporating a paper by the author of this book in *Curating the Great War*, which I co-edited with Nicolas J. Saunders.[1] In it the author adumbrated many of the themes that receive much fuller consideration over the coming pages. The paper was of particular interest to me because of my own long curatorial career in the Imperial War Museum, now IWM. I was naturally familiar with the existing literature on the history of IWM, with its heavy emphasis on the extraordinary nature of the museum's creation and the pioneering character of its early collecting activities. Consequently, my curiosity was immediately piqued by a study that promised to break new ground by focussing on the impact of the Second World War on the IWM. This book has redoubled my interest.

A happy consequence of my long association with IWM is that it affords me the opportunity to enliven this foreword with some personal perspectives. I joined the staff of IWM London in 1989 and it struck me, upon reading this book, that, in structural terms at least, the museum then was more reminiscent of its inter-war self, as described herein, than of the institution we see today. It retained a relatively flat organisational structure. As a junior curator, only the Keeper of my department stood between me and the Director-General. And many of its staff held posts which had survived from the museum's inception. There was still a Museum Superintendent, Warders in blue uniforms with caps patrolled the premises and (essential back then) there was a typing pool.

In those days, in common with other museums, many of the employees, particularly the curators, were 'lifers' who had spent years in post and generally did not expect to move outside the institution. I often wondered aloud if the place was, in actuality, still the Bethlem Royal Hospital (which had formerly occupied the museum's building until 1930) and whether we so-called employees were not in fact being institutionalised on the cheap—kept compliant and happy by assurances that we were doing useful work. In this sort of context institutional memories develop, with anecdotal 'history' being passed on from older to younger staff. Specifically, and with much relevance to this book, memories of the directorship of Noble Frankland remained relatively fresh among my older colleagues, with anecdotes about his interactions with his staff still circulating. Nobody, however, had any first-hand recollection of the previous Director-General. Indeed, while Dr Frankland—although himself a man who would punctiliously address his staff members as Mr, Mrs or Miss—was familiarly referred to as 'Noble', no one seemed to even know the first name of 'Mr Bradley'. The latter was remembered as an old gent who when sleeping at the museum would put his milk bottles out on the front portico and hang his laundry to dry in the boardroom.

The belief, later to be enshrined in Dr Frankland's memoir *History at War: The Campaigns of an Historian*, that the Second World War had been viewed with distaste by Mr Bradley, and accepted by him as a probably fatal threat to the entire existence of the IWM, was widely, possibly universally, held among IWM staff.[2] To those of us who worked with the collections, this way of thinking was reinforced by our knowledge that the collecting of exhibits relating to the Second World War had been a more haphazard and iterative process than the extraordinary efforts made to collect directly from the battle fronts (or at least two of them) during the First World War. As the author points out, this conception has been aired repeatedly in publications by IWM staff over the years. The central achievement of this book, in my view, is to provide solid evidence that it was far from the truth.

The Second World War was a pivotal point in the history of IWM. First it caused a lengthy caesura in its public accessibility. Indeed, the wartime closure was the longest in IWM's history—easily exceeding those inherent in redevelopment programmes undertaken in the 1980s and 2010s and the closure during the coronavirus pandemic of 2020–2021. Secondly it had deleterious physical effects in the form of the scattering of IWMs collections and actual damage from aerial attack. Thirdly, the coming of a new war posed an existential question. Uniquely among British cultural institutions of the era, the war directly challenged the museum's terms of reference and *raison d'être*. Thus, while only sparingly resourced, the IWM faced far greater challenges than other museums and galleries, both in recovering from the war and in reinventing itself to stay relevant in the post-war world. As this book reveals, not only did the IWM prove ready and willing to face this test, but began, as early as 1940, to reconfigure its public offer to meet the new circumstances.

To truly understand the dimensions of the task faced by Lesley Bradley as Director-General in the immediate post-war era, it is important to comprehend the physical circumstances of the IWM at the time. The museum stood in a relatively impoverished area of London, largely unfrequented by tourists or other likely museum visitors. The neighbourhood had endured a significant level of destruction by aerial attack during the war. Indeed, one of the most famous evocations of the Austerity era, the Ealing Studios film *Passport to Pimlico*, was filmed largely on a set created on a bombsite on Hercules Road, just a stone's throw from the museum. Furthermore, as the author makes clear, the museum building itself was in need of repair after bomb damage. Moreover, it was simply not large enough to accommodate the new collections that would potentially arrive if the IWM was to carry through its reinvention as a museum of both world wars. Consequently, it was at this juncture that the IWM took its first steps to becoming the multi-site entity that it is today.

The author's examination of the way in which the IWM was piloted through these turbulent waters is underpinned by a welcome willingness to examine the minutiae of institutional structure, funding and staffing. His analysis also places the story within the context of crisis management. To my knowledge nobody else has attempted such an approach. Previous studies have been loaded towards the more visible themes of exhibitions and collecting, rather than the nitty-gritty of *how* the IWM actually functioned as an institution. In my experience academics are all too frequently eager to ignore such structural and functional considerations in favour of attempting to explain outcomes by detecting, or even inventing, institutional philosophies.[3] Such a methodology necessitates acceptance of the highly erroneous supposition that collecting and exhibition-making take place in a spatial, temporal and financial void. It is, I am happy to record, an approach rejected in this book.

So, despite extensive reading on the history of IWM, and my 32 years of service in the museum itself, this book has been a revelation to me. It is a work that fills a gap in our collective memories and performs an important service in giving due weight to an era of rapid change and redefinition that does not deserve to be obscured from our view by the shinier redevelopments of the mid-1960s, late-1980s and 2010s.

London, UK Paul Cornish

Notes

1. Philip W. Deans, 'Curating the Great War at the Imperial War Museum during the Second World War', in *Curating the Great War*, ed Paul Cornish and Nicholas J. Saunders (Abingdon: Routledge, 2022), 86–104.
2. Noble Frankland, *History at War: The Campaigns of an Historian* (London: Giles de la Mare, 1998), 163–164.
3. Paul Cornish, 'Introduction', in *Curating the Great War*, ed Paul Cornish and Nicholas J. Saunders (Abingdon: Routledge, 2022), 2.

Acknowledgments

This book has been ten years in development. Its genesis can be traced back to my early postgraduate studies at Newcastle University in 2013, where I began thinking critically about the museum known today as IWM (an acronym for Imperial War Museums) while undertaking a master's degree in museum studies. Here, I wrote my dissertation on the Imperial War Museum (as IWM was formally known) during the Second World War. Through this project, I sought to understand how the Second World War impacted the Imperial War Museum in various ways. Though I did not know at the time, that dissertation would end up comprising a pilot study for what eventually became my doctoral research project investigating crisis management at the Imperial War Museum during the Second World War, the research on which the study presented herein is based. The latter project was also undertaken at Newcastle University and spanned 2015–2022, competing for attention with work placements, employment and other life events, plus a global pandemic.

Throughout this journey, I have benefited from the support and encouragement of many different individuals and organisations, all of which are due my thanks. In the first instance, I want to thank the Arts and Humanities Research Council, the Northern Bridge Doctoral Training Partnership and Newcastle University for supporting the original doctoral research project. Administered via the Northern Bridge Doctoral Training Partnership at Newcastle University, the Arts and Humanities Research Council provided three-and-a-half years of research funding. Newcastle University then provided additional support until the project was finally completed. Without this support, completing the following research would not have been possible. I am, therefore, extremely grateful to and humbled by the many individuals from these institutions who believed and invested in me and my ideas.

At Newcastle University specifically, I express particular thanks to my doctoral research supervisors, Rhiannon Mason and Susannah Eckersley. Their guidance throughout that process ensured the original thesis remained 'on track' until the end and was completed in good order. The mentorship they provided has enabled me to grow and mature as an academic. Susannah also receives thanks for her part towards inspiring my doctorate by being the supervisor of the pilot master's dissertation and, thus, for putting up with me a second time. I also thank my colleagues past and present in Media, Culture, Heritage at Newcastle University for their support and encouragement over this research journey. It is a privilege to have learned, worked with and discussed ideas among such engaged and engaging individuals over the past eight years. I cannot conceive of a richer, more diverse body of expertise to draw on while undertaking this research and then converting it for publication.

Away from the academy, this project would have been impossible to accomplish in its final form without the support received from archival staff at IWM and The National Archives. Being a historical study, my methodology relied heavily on me consulting many and varied documentary sources. The work of others in facilitating this was considerable. Indeed, my data folders for IWM alone indicate that I consulted more than 850 accessioned files, exceeding 1500 individual documents. Each file needed to be located, prepped and delivered to the reading room by somebody from the archival team. This consumed considerable time and effort on their parts, so I am incredibly thankful for their efforts. Interest and support were also gratefully received from many other staff at IWM. Particularly, I want to thank Carl Warner and Suzanne Bardgett, who provided opportunities for me to embed myself in the museum community and to contribute my research to its work. This facilitated further valuable discussion and refinement of the ideas presented herein.

Of course, I must thank my peer-reviewer, whoever they are, for their generous remarks on and ideas towards developing this work. I also thank my editorial team at Palgrave Macmillan for commissioning the book and assisting me in bringing it to fruition.

Finally, I close with wholehearted thanks to my wife, family and friends for their personal unceasing support and encouragement while this book came about. It has been a long, challenging and occasionally arduous trek, but it was made enjoyable by those whom I hold dear walking the path with me at different times and in different capacities: Sophie (my guiding star); Mum, Dad, Douglas, Owen and Nanny; the Gurnett 'massive'; Alix, David, Erica, Marloes, Jess and the rest of the 2.69 crew; Mike and Aanchal; the 'Griffins'; Stuart and Chris; Clare; Amelia; and Aron, amongst others. There are, unsurprisingly, many other individuals whom I should acknowledge for their support but am unable to here. I am, nonetheless, exceedingly grateful to each and every one of them and recognise that this book exists through assistance received from many more individuals than could ever be mentioned here. Each has my thanks. You know who you are; I am grateful for what you did.

Contents

About the Author

Philip W. Deans is currently an Associate Lecturer at Newcastle University. He has a diverse background which informs his work, including completing studies in performing arts, military history and museum studies. A museologist, historian and crisis researcher, his work has focussed on the ways that museums and other cultural and creative organisations have managed and resolved crisis-conducive situations. From 2015 to 2018, he was Secretariat for the UK National Committee of the Blue Shield, supporting its campaign to get the British government to ratify the 1954 *Hague Convention for the Protection of Cultural Property in Armed Conflict*. Since 2016, he has sat on the board of the Museums and Galleries History Group, assisting with the group's management and programming.

List of Figures

List of Tables

CHAPTER 1

Introduction

A History of a Museum, An Investigation into Museum Crises

The following study is about museum crises. It explores how crises can disrupt museums and the contrasting defensive and revolutionary approaches to crisis management which museums might take when resolving crises. This investigation is carried out through a historical case study into the Imperial War Museum, London, during the Second World War era, 1933–1950, with particular focus on the war years themselves, 1939–1945. Hitherto, the Second World War history of this important national museum has been under-represented in academic work. The extant historiography comprises only a few studies which limitedly consider the subject. This is surprising, for the years of 1939–1945 were years wherein the museum reinvented and assumed an alternative paradigm with lasting institutional consequences. Conceived as the National War Museum at the height of the First World War, 1914–1918, it eventually opened in 1920 as the Imperial War Museum.[1] In 2011, the museum was rebranded as IWM: an acronym for Imperial War Museums.[2] This marketing exercise acknowledged the museum's now multi-sited nature, for between 1969 and 2002, it acquired four additional satellite sites.

At present, IWM comprises five museums and heritage attractions. Through these, it aspires to be 'a global authority on conflict and its impact', aiming to 'help people understand why we go to war and the

P. W. Deans, *Crisis, Reinvention and Resilience in Museums*,
https://doi.org/10.1007/978-3-031-67806-6_1

effect that conflict has on people's lives'.[3] In doing so, the museum focusses on the experiences of the United Kingdom, the British Empire and the Commonwealth since 1914.[4] The oldest site is IWM London, Southwark, its headquarters and primary museum about war and armed conflict.[5] The second major site is IWM Duxford, south Cambridgeshire, an aviation-orientated museum and preserved First World War, Second World War and Cold War era military airfield.[6] This facility also serves as the museum's large object store, enabling the collection of aircraft, vehicles and other large and heavy *matériel* for which the museum has become renowned over its history.[7] The third major site is IWM North, Salford, an audio-visual heavy museum and interpretation centre.[8] Alongside these three main sites are two other London-based heritage attractions. They are the Churchill War Rooms, City of Westminster, comprising the preserved Second World War era Cabinet War Rooms and Churchill Museum, and HMS *Belfast*, moored in the Pool of London on the River Thames, a preserved Second World War and Cold War era warship.[9]

This development is the product of previous change in ethos at the museum, change that can be traced back to the turbulent events which arose during the Second World War. Indeed, as will be argued herein, the museum in existence today only does so because internal and external challenges and pressures, brought about by the Second World War, catalysed the need for a wholly new *raison d'être* and rationale, prompting questions considered over the coming chapters about the propensity of museums to overcome crises.

The reason that the Imperial War Museum has been selected as the focus of this study is because it provides an opportunity to better understand the concept of crisis, whether different typologies of crisis can be associated with particular states which museums, indeed any organisation, find themselves in, and the interplay between museums and different typologies of crisis. The study considers several practical and theoretical crises over subsequent chapters that the Imperial War Museum arguably faced during the Second World War. The first was a practical crisis of inactivity, caused by wartime disruption to what has been framed herein as the museum's civic service. Hostile conditions, plus local and national war requirements, demanded it adopt special measures for satisfying the wants, needs and interests preoccupying core users and the wider public. The second was a practical crisis of physical destruction, caused by the effects of the war on the Imperial War Museum, specifically of German aerial attacks against London via the Blitz campaigns, 'tip and run' raids and ground-launched V(engeance)-weapon strikes.[10] These attacks produced substantial damage across London, including to cultural institutions. One post-war report on bomb damage sustained by the national museums singled out the Imperial War Museum for its unparalleled damage compared with neighbouring institutions.[11] And the third was a theoretical crisis of irrelevancy, caused by the new social and cultural context that the war ushered in, creating an unenvisaged context which the Imperial War Museum had never been intended for. Until the Second World War, the previously occurring First World War (formerly known amongst Anglophone speakers as the Great War), had been popularly understood as the 'war to end all war': a sentiment often attributed to Woodrow Wilson, but which originates from a thesis conceived and published by H. G. Wells titled *The War That Will End War*.[12] The Imperial War Museum operated on this assumption for its first 22 years.[13] But the Second World War overturned that understanding, in turn establishing a future of uncertainty.

Research Themes, Approaches and Motivations

Conceptual Strand

In exploring the capacity and capability of museums at managing crises, this study takes a deliberately broad view over the concept. It considers crisis and crisis management from different angles. It also considers concepts that feed into the discourse and understanding of crisis management, including attributes required when undertaking crisis management, and strategies which can be deployed to achieve this. As such, the study not only considers crisis emerging from structural or systemic degradation, but also societal change. The case of the Imperial War Museum during the Second World War is an ideal one to investigate these variances in crisis. As discussed, it experienced danger from German aerial attacks against London. And it also experienced the threat of becoming irrelevant in the post-Second World War era if it did not reinvent itself conceptually to take into consideration this new conflict. All the situations faced are considered over Chaps. 4 through 8.

During each of these chapters, a new concept is analysed and examined against the case study alongside previously introduced concepts. Concepts are valuable academic tools in historical organisational research. They support researchers to conceptualise connections, relationships and greater meaning between different events.[14] For example, the concept of crisis can reveal ways in which various difficulties at the Imperial War Museum cohered and formed critical existential phenomena there. It can also help ascertain each phenomenon's significance. Accordingly, concepts formalise the interpretation of information derived from primary sources. They also profile the working assumptions made about this information during the research process.[15] The three main conceptual lenses deployed herein are crisis, reinvention and resilience. In using these concepts, the study shows that the Imperial War Museum was confronted during the Second World War era with systemic crises which the institution had to manage its way either out of or away from.

It is an inevitability that museums will at some point encounter a crisis challenge to their continuance, both crises possessing defensive and revolutionary properties.[16] For some, the challenge results in the museum's closure. For others, the challenge is successfully overcome. Whether a museum experiences the first or second outcome in part depends on the approach taken to the situation's management. Hence, it remains the proposition of this study that museums will prevail or fail depending on two key considerations. The first is their readiness to overcome adversity. In this context, 'readiness' means the wherewithal to confront, resolve and survive crises. That wherewithal could comprise the development of strong and sturdy physical infrastructure and supple *raison d'être* and rationale. But it might also comprise a diverse, qualified and creative workforce; strong, supportive culture; and reliable support networks. By being ready for adversity, museums should be ready for crises—essential if they are to survive. The second consideration is the ability to adjust as and when required. In this context, to 'adjust' means to modify when the prevailing paradigm no longer suits the prevailing environment. No social system inhabits a protracted period of stasis, even if day-to-day life creates this perception. In stable societies, change occurs over generations.[17] Museums must be attuned to this, being, as they are, followers of societal trends.[18] The non-reflexive museum risks becoming left behind by society, rendered irrelevant, and abandoned.

Historical Strand

This study considers the activities of the Imperial War Museum during the Second World War era. As will be shown, the Second World War was an important period in the museum's development. And yet, the case has attracted little critical attention of museum historians or even of the staff there.[19] This raises legitimate questions about whether IWM, hitherto, has fully understood itself in historical and contemporary contexts.[20] The current limited extent of the historiography likely stems from multiple reasons, though a notable possible factor is the existence in the IWM museum archive of an unpublished, internally produced war history of the museum.[21] Comprising two volumes, this institutional autobiography discusses, first-hand, activities at the Imperial War Museum during the Second World War era. Volume one covers the years 1933–1943, volume two 1944–1946. It was commissioned by the War Cabinet in 1942, towards a monograph (which never materialised) on the national museums and galleries during the Second World War, for the state backed multi-volume official war history.[22] This source's existence may have fuelled past perceptions (including those shared anecdotally with the author by IWM staff before research commenced) about the museum's inertia during the war years, and about there being little more to discover on the museum's experience over the conflict. Another possibility stems from the scarcity of photographs depicting the museum during the Second World War and the immediate aftermath, a feature formerly identified by Sue Malvern.[23] Visual material culture can be influential in stimulating historical enquiry.[24] This is especially the case when no other impetus exists. In view of that, the limited pictorial evidence of the activities at the Imperial War Museum during the Second World War era may have compounded an oversight of the institution's wartime history.[25]

Whatever the reason or reasons for the limited scholarship on the Imperial War Museum during the Second World War, the dearth has fostered contestable perspectives amongst interested parties that has proceeded to mythologise the museum's historical existence, with ramifications for contemporary understanding. One example, discussed later, is the proliferation of the narrative told by Noble Frankland, the museum's Director-General from 1960 until 1982. In two publications, Frankland portrays the museum as having lost its purpose during the Second World War, thanks to an ineffectual Director-General who was more content with winding the museum up than working for its lasting continuance.[26] With no other commentary seemingly available, this narrative has been recounted by employees during their own publications on the museum.[27] Such situations are clearly undesirable for any museum, indeed any organisation, for they do little more than strengthen the myth and legend which exists about them to the detriment of knowledge with greater basis in fact.[28]

Taking inspiration from Eilean Hooper-Greenhill's 'effective' (layered) approach to museum history, this study remedies a perceived lack of understanding about the Imperial War Museum over the years spanning 1933–1950, particularly 1939–1945.[29] It does so by researching what John Tosh calls the case's 'historical process'.[30] This was the museum's critical development from a museum on the 'war to end all war' to a museum on both the First World War and the Second World War, all the while fending off the effects of aerial attacks on London and maintaining a civic service. The study argues that the form IWM assumes today derives from decisions and actions taken there to overcome challenging situations posed by the Second World War. Specifically, it posits that the museum, which formally reopened after the Second World War, had evolved from the one which closed at the conflict's start. It shows that the course followed on reopening was set towards fulfilling a totally different mission from the mission originally set out on at establishment.

Structure of the Book

This study is presented over seven chapters, not including the current, opening introduction, or its final, closing conclusion. The first two chapters herein establish the conceptual and historical context of the study. Chapter 2 considers the concepts that support its critical analysis and narrative structure. This is a particularly important chapter which sets out the theoretical foundations underlying the study. Chapter 3, the first of the six case study chapters, then considers the Imperial War Museum as an institution, organisation and museum from the institution's foundation in 1917 to 1939 and the Second World War. It reveals that even before the Second World War, the museum was familiar with making adaptions which counteracted external environmental pressures. It also shows that the museum of the Second World War era possessed institutional, organisational and museological components consistent with many institutions, organisations and museums today, highlighting this study's contemporary relevance to museum practice and research.

The next two chapters address the aforementioned practical crises facing the Imperial War Museum during the Second World War era as well as the mitigating responses that it applied against them. Chapter 4 considers the crisis of inactivity caused by disruption to the museum's civic service provision throughout the conflict. This involves exploring and understanding how organisational resilience facilitates successful museum crisis management, issues and ideas visited further during each subsequent chapter. Chapter 5 then considers the crisis of physical destruction caused by German aerial attacks against London on the museum's collections, building and other critical assets and facilities. It argues that crises which cause physical disruption and structural degradation to museums, such as those arising from deliberate acts or natural disasters, require a defensive response.

And the final three chapters address the abovementioned concurrent theoretical crisis facing the Imperial War Museum and, again, the mitigating response. Together, they consider the crisis of irrelevancy, caused by the new social and cultural context that the war ushered in. Chapter 6 focusses on the work of sensing this threat, conceptualising a suitable response and initiating the process of responding, whereas Chaps. 7 and 8 focus on the work of delivering the response. The chapters argue that crises which cause non-physical disruption to museums, such as those arising from societal value shifts, should be resolved by the museums accepting revolutionary institutional change that recoheres it with society.

Data Collection and Analysis

This study is predominantly founded on archival material researched at the IWM museum archive and The National Archives of the United Kingdom over concentrated research trips in the academic years of 2016–2019. Archival sources encompass heterogeneous documents conveying human thought and expression about occurrences from the everyday to one-off events, with museum archives being no exception.[31] Accordingly, many different documents were consulted and have been used herein, the most common being letters and other written correspondence, meeting minutes, reports and accounts. Each document drawn on herein will have been produced and preserved through a unique mesh of stimuli, systematic processes and personal or institutional agendas. It follows that every document consulted possessed different meanings and informativeness, requiring careful analysis to be understood.

Analysis of the archival and other primary source material occurred alongside the collection process and continued into 2019–2020. As with all historical case study research, the aim of the analysis was to seek to understand the information obtained and to critically evaluate it. Hence, the critical approach finally deployed comprised a three-stage process of critical evaluation. The first involved inspecting the sources under analysis to confirm their authenticity.[32] The second involved critically reviewing the information they presented to determine its validity.[33] And the third involved assessing what the data implied, how it correlated with other units of data, and its significance.[34] The process can be best described as an interrogation of a researcher's assumptions. It involves anatomising and reassembling the extant data in ways that make the information more knowable and meaningful. The result is an ability to make critical observations which inform the understanding of the case.[35]

It should be noted that many of the documents analysed and cited herein originate from or pertain to one man, the Director-General of the Imperial War Museum during the Second World War era, Leslie Ripley Bradley. Alongside historical (including local) archiving practices, this feature results from Bradley's dominance at the museum.[36] The Imperial War Museum was a relatively small institution during the Second World War era, employing around 15 back-of-house staff. By necessity, this study reflects Bradley's omnipresence in the museum archive and case analysis, about which more later.

Use of Acronyms

Finally, it should be noted that even before 2011 the Imperial War Museum was often known and referred to as 'the IWM'. This acronym is pervasive throughout the archival sources cited herein. To avoid confusion during discussion on the museum in historical and contemporary contexts therefore, the use of acronyms has been avoided in the main text unless under the following two circumstances: where they are directly quoted from a source, and where the name of an entity is officially represented by one. Consequently, the acronym IWM, when used outside quotation marks (as done so here), concerns the museum from 2011 onwards.

Conclusion

This historical institutional museum study explores the Imperial War Museum during the Second World War and the years either side: a study of institutional endurance, from which lessons can be derived and applied in the present context. Presented over seven chapters concerning issues framed around crisis, reinvention and resilience, it investigates the historical process through which the museum transformed over 1939–1945 from a museum on the 'war to end all war' to a museum on both the First

World War and the Second World War (the 'two world wars'). In doing so, the study reveals possibilities and options for museums today facing hostile conditions or major change in their constituencies or users' needs.

The study brings sharply into focus the potential adaptability of museums generally when faced with critical existential situations and/or operationally difficult environments. It comprises what Tosh calls a 'knowledge bank', effectively presenting various different possibilities and options derived from the experience of the Imperial War Museum during the Second World War and surrounding years. This is something which contemporary policy makers can draw on when planning for existential difficulty.[37] The findings exemplify a museum's ability—that of their staff—to evaluate situations being confronted and implement appropriate mitigating strategies. Together, these can secure a museum's short-term and long-term survival. By focussing on the Imperial War Museum during the Second World War and surrounding years, the research reveals and emphasises the resilience and creativity emanating from museums in operationally difficult environments. It also demonstrates how such resilience and creativity might manifest through any such organisation's actions.

Notes

1. Gaynor Kavanagh, *Museums and the First World War: A Social History* (London: Leicester University Press, 1994), 117, 146.
2. IWM, 'IWM Launches New Brand', https://www.iwm.org.uk/sites/default/files/press-release/New_Brand_Press_Release.pdf (18 March 2024).
3. IWM, 'Annual Report and Accounts, 2022–2023' ([London?]: HMSO, 2023), 6, https://assets.publishing.service.gov.uk/media/64db55ea3fde6100134a525b/Imperial_War_Museum_Annual_Report_and_Accounts_2022_to_2023.pdf (18 March 2024).
4. Stephen Jaeger, *The Second World War in the Twenty-First-Century Museum: From Narrative, Memory and Experience to Experientiality* (Berlin, BE: Walter de Gruyter, 2020), 227.
5. IWM, 'IWM London', https://www.iwm.org.uk/visits/iwm-london (18 March 2024).
6. IWM, 'IWM Duxford', https://www.iwm.org.uk/visits/iwm-duxford (18 March 2024).
7. Paul Cornish, 'Extremes of Collecting at the Imperial War Museum, 1917–2009: Struggles with the Large and the Ephemeral', in *Extreme*

Collecting: Challenging Practices for 21st Century Museums, ed Graeme Were and J. C. H. King (New York, NY: Berghahn, 2012), 161.
8. IWM, 'IWM North', https://www.iwm.org.uk/visits/iwm-north (18 March 2024).
9. IWM, 'Churchill War Rooms', https://www.iwm.org.uk/visits/churchill-war-rooms (18 March 2024); IWM, HMS Belfast', https://www.iwm.org.uk/visits/hms-belfast (18 March 2024).
10. Richard Overy, *The Bombing War: Europe, 1939–1945*, pbk edn (London: Penguin, 2014), 73, 120; Chris Goss, with Peter Cornwell and Bernd Rauchbach, *Luftwaffe Fighter-Bombers Over Britain: The Tip and Run Campaign, 1942–1943* (Mechanicsburg, PA: Stackpole, 2003); Christie Campbell, *Target London: Under Attack from the V-weapons During WWII* (London: Abacus, 2013).
11. Standing Commission on Museums and Galleries, *Third Report: The War Years and After* (London: HMSO, 1948), 17.
12. H. G. Wells, *The War That Will End War* (London: F. and C. Palmer, 1914); David Budgen, 'Literature' in *1914–1918-online. International Encyclopedia of the First World War*, ed Ute Daniel, Peter Gatrell, Oliver Janz, Heather Jones, Jennifer Keene, Alan Kramer and Bill Nasson (Berlin, BE: Freie Universität Berlin, 2014), 4, 10.15463/ie1418.10329/1.1, https://encyclopedia.1914-1918-online.net/pdf/1914-1918-Online-literature-2018-06-29-V1.1.pdf (18 March 2024).
13. Sue Malvern, 'War, Memory and Museums: Art and Artefact in the Imperial War Museum', *History Workshop Journal* 49 (2000): 182–193.
14. Mairi Maclean, Charles Harvey and Stewart R. Clegg, 'Conceptualizing Historical Organizational Studies', *Academy of Management Review* 41 (2016): 624.
15. Mary Fulbrook, *Historical Theory* (London: Routledge, 2002).
16. Elaine Heumann Gurian, 'Preface', in *Institutional Trauma: Major Change in Museums and Its Effect on Staff*, ed Elaine Heumann Gurian (Washington, DC: American Association of Museums, 1995), 17.
17. Montgomery Van Wart, 'The First Step in the Reinvention Process: Assessment', *Public Administration Review* 55 (1995): 429.
18. Graham Black, 'Introduction: The Challenge of Change', in *Museums and the Challenge of Change: Old Institutions in a New World*, ed Graham Black (Abingdon: Routledge, 2021), 3.
19. The three most significant contributions include N. J. McCamley, *Saving Britain's Art Treasures* (Barnsley: Leo Cooper, 2003); Debbie Whittaker, 'Museums at War: The Impact of the Second World War on the National Museums of London, with Particular Reference to the Imperial War Museum, National Gallery, Victoria and Albert Museum and the British Museum' (master's dissertation, University of Leicester, 2010); Caroline

Shenton, *National Treasures: Saving the Nation's Art in World War II*, pbk edn (London: John Murray, 2022).

20. Kate Hill, 'Why Study Museum History?', Blog, Museums and Galleries History Group, https://www.mghg.info/blog/2017/5/27/why-study-museum-history (18 March 2024).
21. IWM, Museum Archive (MA), EN1/1/REP/032, 'War History of the Imperial War Museum, 1933–1943'; IWM, MA, EN1/1/REP/032, 'War History of the Imperial War Museum: Part II, January 1944 to the Reopening of the Museum on 27th November 1946'.
22. IWM, MA, EN2/1/MUS/001/1, 'For the Directors' Conference', 11 November 1942.
23. Malvern, 'War', 194.
24. Leonie Hannan and Sarah Longair, *History Through Material Culture* (Manchester: Manchester University Press, 2017), 45.
25. See Paul Cornish, 'Foreword', this book, for additional possible reasons.
26. Noble Frankland, review of *Museums and the First World War: A Social History*, by Gaynor Kavanagh, *Journal of the History of Collections* 7 (1995): 127; Noble Frankland, *History at War: The Campaigns of an Historian* (London: Giles de la Mare, 1998), 163–164.
27. Terry Charman, '"A Museum of Man's Greatest Lunatic Folly": The Imperial War Museum and its Commemoration of the Great War, 1917–2008', in *A Part of History: Aspects of the British Experience of the Western Front* (London: Continuum, 2008), 104–105; James Taylor, 'Interpreting the Second World War', in *The Power of Objects: Museums and World War II*, ed Esben Kjeldbæk (Edinburgh: MuseumsEtc, 2009), 55–56.
28. Yinnis Gabriel, *Storytelling in Organizations: Facts, Fictions, and Fantasies* (Oxford: Oxford University Press, 2000).
29. Eilean Hooper-Greenhill, *Museums and the Shaping of Knowledge* (London: Routledge, 1992), 9–12.
30. John Tosh, *The Pursuit of History: Aims Methods and New Directions in the Study of Modern History*, rev. 3rd edn (Harlow: Longman, 2002), 38.
31. Sammie L. Morris, 'Starting from Scratch: How to Create a Museum Archives', *Libraries Research Publications* 45: https://docs.lib.purdue.edu/cgi/viewcontent.cgi?article=1052&context=lib_research (18 March 2024).
32. Michael Stanford, *A Companion to the Study of History* (Oxford: Blackwell, 1994), 154.
33. Robert K. Yin, *Case Study Research: Design and Methods*, 5th edn (Thousand Oakes, CA: SAGE, 2014), 118–123.
34. Robert E. Stake, *The Art of Case Study Research* (Thousand Oaks, CA: SAGE, 1995), 71.
35. Ibid., 75–77.

36. Aleida Assmann, 'Canon and Archive', in *The Collective Memory Reader*, ed Jeffrey K. Olick, Vered Vinitzky-Seroussi and Daniel Levy (New York, NY: Oxford University Press, 2011), 337.
37. Tosh, *Pursuit of History*, 30; Diane L. Douglas, 'Historical Analogies, Use of', in *Encyclopedia of Crisis Management*, ed K. Bradley Penuel, Matt Statler and Ryan Hagen (Thousand Oaks, CA: SAGE, 2013), 469.

CHAPTER 2

Key Framing Concepts

Introduction

Underpinning every study on some complex sociological phenomenon is a web or patchwork of concepts which supports its production and reading. Indeed, the academic and practical fields that comprise museum studies and crisis management are informed by heterogeneous issues and ideas, demonstrated through publications such as Rhiannon Mason's voluminous edited compendium *Museum Studies* and Christer Pursiainen's cross-disciplinary, non-context-specific handbook *The Crisis Management Cycle*.[1] This study proactively engages with concepts, which are explicitly used to interpret the findings presented.

The current chapter introduces and explores eight concepts which have helped frame this study on the Imperial War Museum during the Second World War era and interpret its findings. These framing concepts are 'institutions', 'organisations' and 'museums'; 'legitimacy'; 'crisis' and 'crisis management'; 'resilience'; and 'reinvention'. As expressed in Chap. 1, concepts can help to identify connections, relationships and meaning between events, revealing trends and patterns that are not immediately apparent. They can also help to structure complex analyses around shared modes of understanding and communicate them via established scholarly discourses, aiding their recognition amongst readers. In using framing concepts, this study uses what David A. Snow et al. call 'frame

P. W. Deans, *Crisis, Reinvention and Resilience in Museums*,
https://doi.org/10.1007/978-3-031-67806-6_2

amplification' and 'frame bridging'. The former establishes an 'interpretive frame that bears on a particular issue, problem or set of events'.[2] This clarifies 'the meaning of events and their connection to one's immediate life situation'.[3] The latter involves establishing the 'linkage of two or more ideologically congruent but structurally unconnected frames regarding a particular issue or problem'.[4] This serves to form linkages between different but relatable issues and ideas.

The eight aforementioned framing concepts are addressed below over six substantive sections. The first section introduces the concepts of the institution, organisation and museum respectively, three constructs which characterise this study's case. It expounds each, discusses the interlinkages between the three, and considers how institutionalism underlies organisation and how organisation underlies museums. The second section then introduces the concept of legitimacy, arguably the most fundamental 'base-concept' to consider when researching museums and society. It profiles legitimacy and offers illustrative examples in the museum context. After that, the third and fourth sections introduce the concepts of crisis and crisis management respectively. The former elucidates crisis as an abstract concept through contextualising the term's etymology, characterising different ways the concept can be read, examining the phenomenon in both theory and practice, defining the term, and establishing the concept's historicisation—its use today by academics and commentators necessitating this detail. The latter then narrows the focus down to elucidate crisis management, including evaluating various analytical crisis management frameworks and identifying the one used herein. Next, the fifth section introduces the concept of resilience, an important organisational attribute essential when undertaking crisis management in museums. It presents resilience and then relates resilience to the concept of crisis management. And finally, the sixth section introduces the concept of reinvention, a strategy in museum crisis management which can be applied under certain circumstances. It explains reinvention and outlines the approach used herein to critique institutional reinvention processes.

Institutions, Organisations and Museums

Institutions

This study centres on something known as an 'institution'. When considered in abstract, an institution is a complex and dynamic concept which can take manifold forms and belies easy definition.[5] It comprises some physical or non-physical construct created by society to guide, influence or control the way society operates and generates knowledge and understanding. It also perpetuates societal behaviour.[6] Physical examples might include museums, schools and law-courts, whereas non-physical examples might include national and cultural narratives, curricula and legal frameworks. Institutions bring body and definition to the social system by establishing and binding composite elements of a society.[7] As such, they have power over the direction which their respective societies travel. Yet institutions are not all-powerful. They still depend on two factors for their ability to direct social conduct.

The first factor is human interaction. While institutions comprise ideas, words and, for some, even bricks and mortar, it remains only through people's engagement with them that they have potential to impact on the social system.[8] Peter Berger and Thomas Luckmann offer an effective analogy for this dependence: institutions may exist in human conception and physical structures, but they are effectively 'dead' entities unless 'ongoingly brought to life in actual human conduct'.[9] The second factor is societal authorisation.[10] Institutions require approval to set and disseminate the rules and norms of society, a licence implicitly and sometimes only acquiescently given. Without this approval, they become toothless, pointless and ultimately unjustifiable. Just as society cannot function unless energised by institutions, institutions cannot function unless permitted to do so by society.[11]

Organisations

The institution centred on in this study manifests as an 'organisation'. An organisation consists of a structured, systemic entity created and resourced by people to realise aims and objectives in a social system embodying institutionalised cultural beliefs and practices. It can vary in size, purview and manifestation, but typically maintains an instrumental logic.[12] Distinguishing and differentiating organisations from institutions has long been a subject of debate amongst sociologists and management scholars.[13] Some argue that overlap does exist, others that they are mutually exclusive. One helpful intervention in this discussion is made by Elias L. Khalil, who directs attention to the entity's location and spread on the ends-means spectrum. Khalil explains how organisations are defined by their ends. Institutions, conversely, are defined by the means they provide.[14]

This proposes that organisations can be understood as entities which perform specific, preconceived purposes, and institutions as constructs which create the conditions wherein organisational rationale is realised.

Organisations need to ensure that their aims, objectives, performances and behaviours are consistent with the values of the social system in which they operate.[15] As human need modifies following environmental and social change, organisations must adapt their offer. Those which cannot meet human need become redundant and invariably collapse in time.[16] Yet, just as organisations are potentially impacted by changes to the social system, so too is the social system and lived environment potentially impacted by organisations. Take the cost of residential property, for example. One piece of research has found that museums can increase local purchase and rental prices, with broader inflationary consequences.[17] Another has found that abattoirs have the opposite effect.[18] Hence, the societal-organisational power projection does not necessarily flow in one direction only. As with institutions, organisations are at once driven by and drivers of proceedings and attributes constituting the social system.

Organisations in the Institutional Context

Alongside the debate over the distinction between organisations and institutions, there is also debate over the issue of where and how organisations exist in the institutional context. W. Richard Scott identifies three different viewpoints from existing research.[19] The first, posited by Douglas North, advances the idea that organisations are institutionalised agents in a social system created and dictated by external institutional constructs.[20] The second, posited by Philip Selznick and Oliver E. Williamson, advances the idea that organisations themselves, conversely, are institutional actors arising from the decision making of their organisational agents.[21] Both these viewpoints make important contributions to the discourse surrounding the relationships between organisations and institutionalism and offer compelling arguments.

Yet according to other research, the ideas posited by North, Selznick and Williamson only tell half the story. In practice, as Staffan Furusten demonstrates through discussing the transmission of institutional ideas from individuals and organisations to other organisations, organisations adapt with and influence the institutional environment.[22] This falls under the third and seemingly most rounded viewpoint posited by researchers such as John W. Meyer and Brian Rowan, Lynne G. Zucker and Frank R. Dobbin, which advances the idea that organisations constitute both an institutional and institutionalised entity, where its structure and operation is influenced and also geared towards sustaining the cultural beliefs pervading the lived environment.[23] In the words of John Child: 'We always have to bear in mind that while the organization we have reflects our wider society, organization in turn also shapes the kind of society in which we live'.[24] Accordingly, organisations should not be considered entirely free actors. Yet, neither should they be considered agents deprived of all autonomy. Organisations are free to operate and impact on the social system, but with constraints. These constraints are society's norms, rules and beliefs that create the conditions of the social system: conditions which organisations must accommodate in their structure, procedures and goals. As such, it is indeed entirely conceivable that an organisation could also comprise an institution, but one with greater restrictions and limitations on them than an intangible institution. In any case, this debate undoubtedly has many more miles to run.[25]

Museums

The organisation at the centre of this study is a 'museum'. A museum comprises an organisation that, in principle, collects, preserves and/or presents and interprets the cultural heritage of people as well as the natural world. It can be counted amongst the most impactful institutions developed by civil societies and has become or started becoming a regular feature of the landscape across the developed world.[26] In the United Kingdom for example, many cities, towns and even villages possess some museum-esque institution on the local area or pertaining to some other non-geographically specific subject. Museums inform, educate and entertain people. They are also a means for individual and communal definition and representation. Through undertaking this work, museums project ideas and ideologies, take sides over issues, and convey messages both domestically and internationally.[27] And their impact is not purely cultural. It extends to social and economic matters too, as previously indicated. Although occurring outwith their traditional purview, research shows convincingly that museums can be tools for urban regeneration and fiscal growth in deprived communities.[28] The relocation of the Imperial War Museum to the slums of Lambeth just south of the River Thames over 1935–1936 provides an early if unintended example of this.[29]

There has been considerable disagreement amongst museum professionals, the visiting public and academics over recent decades about how to characterise museums today beyond the basic description just provided. Thus, many museum associations and bodies produce and draw on their own descriptors.[30] This is because, arguably, there no longer exists an archetype of museum that fits all national, regional and subject contexts (if one ever truly existed). Rather, museums have become diverse and divergent institutions which take various forms in broad and loose discursive parameters.[31] Through developing from private collections comprising physical 'curios' to public facilities preserving and presenting physical and intangible collections, as well as historical buildings and landscapes, museums are now heterogeneous entities.[32] The fierce rejection of one particular broad-based definition for the museum in 2019, proposed by the International Council of Museums, exemplifies this situation.[33] It also demonstrates what David Lowenthal identified ten years prior as emerging unease and doubt over what museums are and are for with their ever-greater diversification.[34]

To attain such heterogeneity, many museums, particularly from Anglophone countries, have undergone what Kenneth Hudson calls 'a revolution [...] in museum philosophy and in its practical applications'.[35] This derives from two overlapping contexts. One is the established requirement for museums to make incremental changes that ensure they remain aligned with the social system, as described by Graham Black.[36] The other is what Simon J. Knell, Suzanne MacLeod and Sheila Watson describe as a revolutionary predisposition pervading museums, where museum staff reject their institution's projected *status quo* in favour of alternative and more desirable futures.[37] Indeed, according to Knell, MacLeod and Watson, there can be no certainty to museum development. The concept exists in an unpredictable state of flux. Consequently, its historical course offers few insights about the prospects for museum futures other than by demonstrating their tenaciousness at seeking out new sustainable models and, ultimately, support to continue existing and serving society.

Legitimacy

One recurring theme in the previous section, indeed this study throughout, is the societal concern that entities such as institutions, organisations and museums which impact a social system should exist congruently with their respective society's expectations, values and norms. As discussed, institutions need human interaction and authorisation to direct societal conduct. And organisations, including museums, need valuation and patronage via clientele, visitor-bases, constituencies, *etcetera*, to pursue aims and objectives. Put simply, institutions, organisations and museums must satisfy the communities sustaining them that they are worth supporting with intellectual, financial and material resources—that they are 'legitimate'.

The concept of legitimacy is constituted of various issues and ideas that help explain the state of condition, process or perception which some entity must reach to garner societal support.[38] An often-cited definition for it, produced by Mark C. Suchman, describes the phenomenon as a 'generalized perception or assumption that the actions of an entity are desirable, proper or appropriate within some socially constructed system of norms, values, beliefs and definitions'.[39] All institutions and organisations depend on such perceptions of legitimacy, museums being no exception.[40] Over their history, they have been held by proponents and advocates as forces for personal or societal good.[41] Examples of this include, amongst others, conveying prestige amongst elite circles during the eighteenth and nineteenth century, educating populations on their civic responsibilities during the early twentieth century, and dispensing non-museum public services during the early twenty-first century.[42]

Alongside supporting the continuance of museums in a finite social-environmental system, legitimacy can also stimulate forces which support their existence when they become threatened.[43] Indeed, the twentieth and twenty-first centuries offer numerous examples where proponents and advocates have sought to prevent museums from being eliminated. One specific example is the Detroit Institute of Arts in Michigan, United States. This well-established museum became threatened with closure due to unsustainable funding cuts during 2013 when the city of Detroit filed for bankruptcy. Local residents however perceived sufficient value in retaining the museum that they purchased it from the city authorities, thus preventing the museum's liquidation.[44] A second, more general example is the numerous instances of work undertaken by institutional, local and national authorities to protect museums from natural disaster, theft and general degradation.[45] And another general example is the creation and ratification of international humanitarian law by national governments to protect museums and other cultural property during armed conflict.[46]

Legitimacy is a convenient and powerful concept that can offer an effective explanation for why societies accommodate some things and reject others. But its ensuing widespread application, as well as some misapplication, across different disciplines has also resulted in concerns arising over the basic construct becoming obscured by superfluous meaning.[47] Consequently, to avoid unintentionally and unnecessarily exacerbating the problem, a deep and sustained engagement with legitimacy is deliberately avoided herein. This is possible because the principle driving concepts that will be discussed and applied below are sufficiently developed, rounded and capable of supporting the research presented over subsequent chapters. Nonetheless, the concept still features in following chapters where its application leads to insight, demonstrating the concept's significance and helpfulness.

Crisis

Benefits and Challenges of Drawing on Crisis in Scholarship

The first principle driving concept used in this study is the concept of 'crisis'. It concerns a situation wherein normal routines, operations and existences become challenged to the point that they can no longer be performed or maintained effectively, causing serious disruption. This is a phenomenon which in part characterises the case at the centre of the study.

Crisis is a complex concept but one that has also become semantically impoverished through recurrent expansion in meaning over many years.[48] Possessing antiquarian origins, its application has increased significantly since the early twentieth century, with the worldwide Great Recession of 2007–2012 finally relegating the term to an everyday and often expressed word.[49] Notwithstanding specific technical uses in economics, medicine and psychology, crisis is often raised during media and academic discourse as a byword for negative and complex social, economic and political situations.[50] Today, people exploit the term's ability to animate and cohere disparate events, this prompted by the word's omnipresence and facilitated by its powerful rhetoric yet semantic ambiguity. Thus, crisis has become a word that, with little thought, can spice up dull or dry narratives, increasing their intriguingness.[51]

The development and contemporary overuse of crisis might lead some to argue—not unreasonably—that the idea should be avoided as a scholarly framing concept. But abandoning it in academic work would be short-sighted. There is possibility to develop new meanings for the term, which can help produce new semantic value in the word.[52] Firstly, its origins and technical uses still point towards a useful humanities and social science concept that can help researchers analyse critical situations effecting systems. And secondly, where theorised and utilised appropriately, the concept can help researchers pathologise the tectonic undercurrents which cause critical situations in the first place.[53] These include power relations, contestations and conflicts between individuals, communities and social structures.

Defensive and Revolutionary Meanings of Crisis

Over its etymological development, crisis has acquired signification informed by classical legal, theological and medical origins. This has occurred via a process not dissimilar to cultural 'coding' theorised by Stuart Hall, where people's engagement with an idea, sign or symbol produces shared and often multiple meanings of the relevant object.[54] Initially, two general applications for the word crisis arose, becoming the foundations underlying all subsequent developments. In one use, it emerged to describe something or someone requiring critique and judgment against prevailing knowledge, such as by a judicial, or the God of Abraham, Isaac and Jacob.[55] In the other, it emerged to mark the moment when the body, both organism and politic, arrives at some problematic health condition that is either reversed or which progresses until the body's death.[56] Viewed together, these uses generally signpost situations fitting conceptions about the modern world where actual or metaphorical salvation or damnation, or health or disease and the death of something, is brought into question.

While undergoing its development, two useful if contrasting approaches to reading crisis emerged. The first was a defensive reading. This occurs where the body is seen arbitrarily to inhabit the condition of health or sickness. Brian Milstein raises the message in Thomas Hobbes' *Leviathan* as an example (even though Hobbes himself never used the term): 'restricted to the guideposts of a "healthy" or "sick" commonwealth, the political imagination of the Leviathan remained limited to the oppositions between peace and war, stability and chaos, perseverance and decay'.[57] But when Hobbes was writing *Leviathan*, the idea that society comprised a self-aware entity capable of achieving progress started to emerge (including in Hobbes' ideas), influencing the way crisis would later be understood.

Around one hundred years later, crisis also began representing a breaking point in the *status quo* and a transitional phase between two states. Not simple conditions of sickness and health, but rather completely different paradigms of existence.[58] Such ideas can be seen at this time in multiple revolutionary-inspired essays by American philosopher Thomas Paine, each titled *The American Crisis*, written during the American War of Independence:

> By perseverance and fortitude, we have the prospect of a glorious issue; by cowardice and submission, the sad choice of a variety of evils—a ravaged country—a depopulated city—habitations without safety, and slavery without hope—our homes turned into barracks and bawdy-houses for Hessians, and a future race to provide for, whose fathers we shall doubt of.[59]

This latter reading of crisis draws associations with the concept of 'modernity', an approach to understanding the world that rests heavily on the presumption of continual progress of society.[60] Accordingly, crisis also accrued connotations with revolution and liberation: where pressure from stagnation and oppression hits a tipping point and results in the *status quo* being replaced. This understanding informed Karl Marx's ideas on economic crisis. In Marxist thinking, the capitalist system is riddled with contradictions. These inevitably entail the decreasing profitability of industry. To restore profitability, the bourgeoisie owning classes must degrade the proletariat workers' employment conditions. A vicious cycle ensues. Further depreciation prompts further employment degradations, causing mounting pressure at the base of the system. This persists, conceivably until the proletariat's total impoverishment. At that point, the cycle implodes, causing a violent overthrow of the bourgeoisie.[61]

Crisis in Theoretical Understanding

Crises are subjective situations. They occur when museum organisations or museum communities (indeed organisations and communities generally) perceive something disruptive to be impacting them directly or indirectly. Hence, the declaration of a crisis requires much more than a fully theorised definition for the phenomenon. It also requires what Christopher Whitehead et al. call 'a set of interlinked structural and discursive phenomena'[62] which Milstein articulates as 'a range of claims, comments, and attitudes regarding our relation as a collective "us" to a "not-us around us" upon which our everyday life depends'.[63] Thus, to declare a crisis is to take a critical stance on the conditions that something inhabits derived from perceptions about the wider socio-environmental system. An understanding of what crisis *is* therefore requires more than an understanding of what crisis *means*. It also requires an understanding of what crisis *needs* to have social impact.

The nature of that structural and discursive phenomena is established by Milstein over four specific crisis components.[64] They comprise a crisis' 'foundations', 'object', 'resolution' and 'community'. The first component is the 'foundations' on which crisis arises. It stems from a perception of something falling into contradiction with expected or desired norms. This subjectivity can be likened to that surrounding dirt, which Mary Douglas once famously theorised as 'matter out of place'.[65] The second component is the 'object': the thing inhabiting crisis. It could be physical or non-physical but must uphold some aspect of the social system. The third component is the 'resolution' which the crisis conscious believe an object must reach to escape crisis. If no resolution is perceivable, no crisis exists. And the fourth component is the 'community': the people who collectively conceive and spread perceptions of crisis. Communities sustain crises. They identify and strive to rectify whatever discontinuity has alarmed them. Membership can be defined by or outwith geography. It may also cut across established groupings political and social and vary in size depending on the nature and extent of the disruption. Crisis communities arise when somebody declares a crisis which resonates with others. Those others then licence themselves and fellow community members to speak authoritatively about the crisis: its significance and the resolution. This gives the crisis bearing. Ultimately, a crisis' impact depends on the vitality of its community.

Milstein's ideas show that 'crisis' not only potentialises a definition for disruption to systemic entities. It also comprises a concept for making sense about the social system and articulating discomfort arising from developments therein. Hence, crisis represents, on one-hand, symptoms of unpredictability, instability and potential danger, and on the other, an alarm or rallying cry to warn others against those symptoms, preventing them from causing perceived harm. In short, it is a means of analysing and negotiating the social system.

Crisis in Practical Understanding

Crises, once declared, warn of situations in the social system that are disruptive, unstable, unpredictable and potentially dangerous, making for unpleasant experiences to encounter.[66] There are, however, many conceivable occurrences which could cause such unpleasantness, but not all of which will constitute a crisis as theorised above. Rather, some will constitute 'routine emergencies'. Complicating matters, crises and routine emergencies are not fixed to certain typologies of events. It depends on the context. Museums and their communities are susceptible to both crises and routine emergencies. What, in one context, might constitute a routine emergency for a museum or community, could, in another, constitute a crisis. To further understand crises therefore, it is worth setting them apart from routine emergencies.

Crisis scholar Shari R. Veil has identified several key differences between crises and routine emergencies that distinguish the two.[67] These stem from disparity in the regularity and anticipation surrounding them. Routine emergencies are anticipated. They happen regularly, follow familiar patterns and are expected. Accordingly, the respective individuals tasked with their management—the 'crisis respondents'—will have been trained to deal with them proactively following set yet flexible protocols. Michele Wucker has termed these events 'gray rhinos [*sic*]': a metaphor for the big, obvious and not unexpected dangers of the world.[68] Crises, by contrast, are not anticipated. Capable of developing both suddenly with little forewarning or slowly and silently without notice, they happen rarely, have no standard model, and are unexpected. For this reason, crisis respondents will only be able to react to them generally on a back footing. Crises can disrupt the daily routines of museums and their communities in ways which cannot confidently be predicted.[69] This causes an ever-worsening situation: disruption resulting in chaos, chaos resulting in uncertainty, and uncertainty resulting in restricted or uninformed decision-making.[70] Contrasting with Wucker's gray rhinos metaphor, Nassim Nicholas Taleb has termed these events 'black swans': a metaphor for something which is unanticipated, unpredictable, yet devastating on occurrence.[71]

Limited anticipation strikes at the heart of what makes crises potentially so deadly. The museum or community that experiences a danger on a sufficiently regular basis will likely recognise its typical features and know what to expect and how to manage them. Conversely, the museum or community that experiences the same phenomenon but rarely (if ever), or without possessing set management protocols, will likely have little experience of or preparations for dealing with the situation. In this latter instance, when the phenomenon does strike, the museum or community may be caught off guard and susceptible to the effects. Hence, what would usually be considered a routine emergency in the conception of the former museum or community could in the conception of the latter be a crisis.

It should be noted that crises can also develop from routine emergencies. Sometimes, phenomena which typically present as routine emergencies exceed the parameters of their regular manifestation. In doing so they turn into extraordinary events that no museum or community could have reasonably anticipated or prepared for. These extreme events cause unprecedented, non-routine effects on impacted museums or communities, leading to the declaration of crisis. Didier Sornette has termed such events 'dragon-kings': a metaphor for events that are exceptionally large, impactful and unique in origin, exceeding expectations.[72]

Another dimension which distinguishes crises from routine emergencies is their potential protractedness. A museum or community which survive the acute phase of a crisis can experience long, drawn-out effects from the situation. Sometimes, these effects become permanent. There are many nuanced reasons why a crisis might cause such long-lasting or permanent effects. Each instance however can conceivably be distilled down to one of two scenarios. The first is that the crisis management strategy deployed was, for some reason or another, unsuccessful or only partially successful. The second is that the crisis management strategy was indeed successful, but not without causing or accommodating fundamental, lasting change. Routine emergencies, by contrast, are usually short-lived from start to finish, having been anticipated and prepared for with fewer if any ramifications. In turn, the measures and resources necessary for dealing with a crisis tend to be extraordinary. Not just in their protractedness or volume, but also their unprecedentedness or irregularity. Routine emergencies, by contrast, can be more efficiently dealt with, given their familiarity.

Defining Crisis

Considering everything discussed above on the term and concept of crisis—its origins, meaning and implications—'crisis' is defined herein as follows: an uncomfortable, unpredictable, unstable and potentially dangerous situation, where the impacted systemic entity (such as a museum) will be disrupted, perhaps inoperably and irreparably, requiring extraordinary intervention to be overcome. In the organisational context, they are also disruptive situations which defy the expectations of stakeholders and threaten an organisation's central aims and objectives.[73] Bereft of any clear, lasting canons on the term and concept, this definition has been structured around Rüdiger Graf and Konrad H. Jarausch's etymologically coherent recommendation that 'Any viable definition [for crisis] would have to involve assumptions of some kind or another about the normal course of events, an exceptional period of tension in which at least two different outcomes are possible, and a solution'.[74] Yet it disregards their contention that the solution must be 'in the form of a new state of affairs'.[75] This is for two reasons. Firstly, the above discussion about the development of the word has shown that crisis can be read in two ways. To give it a cast-iron revolutionary fixing, which Graf and Jarausch's recommendation would seemingly essentially imply, disregards past defensive readings. Secondly, crises are subjective. This means that not all crises necessarily warrant revolutionary resolutions.

Using Crisis Historiographically

The concept of crisis is often used in historiography to characterise specific historical periods or situations, as occurs with this study. R. J. Overy has previously observed that 'When historians use the word "crisis" they usually employ it with hindsight, taking all the facts together and imposing on them greater coherence or significance than was perceived by contemporaries'.[76] There is nothing wrong with this approach in principle. After all, when carrying out any research, all academics draw on their own interpretations of data to create meaning that bespeaks relevant issues and ideas. Where problems start arising with the approach, however, is when crisis becomes a casual explanation for historical process. As Graf and Jarausch caution: 'Transferring the historical diagnosis of crisis directly into the historiographical narrative [...] can easily be misleading as it totalizes one perspective on the past which was most likely formulated with specific interests'.[77] This is an important point which underscores the subjectivity of crisis discussed above; while one person may perceive some situation to exhibit crisis symptoms, the same perception may not be shared by another: what Max Kölbel terms a 'faultless disagreement'.[78] It follows that crisis concepts need to be handled with consideration when applied historiographically so that, where necessary and appropriate, accommodation can be made for other potential subjective interpretations.

There are two such approaches that historians can take to historicising crisis. The first, devised by Graf and Jarausch, entails discovering how historical actors understood the term at any given time and deployed the word and its meaning when describing their prevailing social, political and economic situation.[79] Where crisis consciousness has been conveyed by primary sources, historians can use the concept on those terms to critique the condition that the relevant entity such as a museum or community held. The second approach, devised for this study, involves discovering whether some entity could be considered to be gripped by crisis. More representative of the approach described by Overy than Graf and Jarausch, it is most appropriate where crisis consciousness has not surfaced from primary sources, as was the case with the primary sources consulted towards the research presented herein.

The second approach involves analysing a hypothetical crisis object to see whether it exhibits what are considered over the following chapters as 'crisis-conducive' conditions. Through doing so, the historian can try at understanding the way that some crisis-conducive situation was resolved. The upshot is an appreciation of how similar declared and legitimised crises are negotiated and what implications they might have for the crisis object—as well as everyone and everything dependent on it. This approach offers real, proven benefit to those individuals who find themselves confronting a crisis. Joanne E. Hale, David P. Hale and Ronald E. Dulek have found that when resolving crises, respondents can obtain assistance from past accounts of crisis management during crisis decision making, even if the situations being faced bear only partial or limited resemblance with the problem at hand.[80]

Crisis Management

The unpredictability and potential volatility of a crisis can make them extremely hazardous to museums and their communities. An object in crisis may deteriorate significantly and perhaps terminally without some organised force counteracting against the effects. This is the work of 'crisis management', a concept and process geared towards preparing for, coping with, and recovering from crisis. The subject drew considerable academic attention during the 1980s following various high-profile man-made disasters which needlessly struck the world throughout that decade, including the 1984 Bhopal gas leak in India, the 1986 Space Shuttle *Challenger* explosion in the United States, the 1986 Chernobyl nuclear meltdown in Ukraine, and the 1989 *Exxon Valdez* oil spill in the United States.[81] A framework for crisis management is used in this study to help structure certain analyses of the case.

Various crisis management frameworks have been developed since the 1980s to help potential crisis respondents to explicate the process. Some focus on the ways that crises manifest in crisis objects. Others focus on the ways that crises materialise.[82] Crisis management frameworks are not crisis management plans. The latter comprise tailored schedules that specify how specific crisis objects should deal with a crisis that has arisen: establishing who does what, the way things should be done, the tools that need to be used, *etcetera*.[83] Rather, crisis management frameworks are tools for understanding how crises impact crisis objects.[84] This study explores museum responses to different crisis-conducive situations by drawing on a crisis management framework that arranges its response over several stages of what Pursiainen calls the 'crisis management cycle'.[85] Table 2.1 outlines four examples from the many that have been developed during the past 30 years.

Table 2.1 Outline of different frameworks for crisis management

The Essential Phases of Crisis	*Three-stage Linear Framework (Richardson, 1994)*	*Four-stage Linear Framework (Fink, 1986)*	*Four-stage, Two-layer Matrix Framework (Crandall, Parnell and Spillan, 2021)*	*Five-stage Linear Framework (Pearson and Mitroff, 1993)*
Before the Crisis	Pre-crisis/ disaster stage	Prodromal crisis stage	Landscape survey stage Strategic planning stage	Signal detection stage Preparation/ Prevention stage
During the Crisis	Crisis impact/ rescue stage	Acute crisis stage Chronic crisis stage	Crisis management stage	Containment and damage control stage
After the Crisis	Recovery/ demise stage	Crisis resolution stage	Organisational learning stage	Business recovery stage Learning stage

Adapted from Crandall, Parnell and Spillan (2021).

The example presented in the second column above is the three-stage, linear variant proposed by Bill Richardson.[86] This represents the most basic of useful crisis management frameworks.[87] It comprises a 'pre-crisis stage', a 'crisis impact stage' and a 'crisis recovery stage'. In this first example, crisis management commences when respondents attempt to prevent materialised crises from effecting relevant crisis objects by eliminating their underlying causes, whatever and wherever they are. On the occasion that prevention proves impossible however, the crisis respondents pivot to mitigation. Once the situation is stabilised, the crisis respondents then move to restore operations and stakeholder confidence.

The example presented in the third column is the four-stage, linear variant proposed by Stephen Fink.[88] It comprises a 'prodromal crisis stage', an 'acute crisis stage', a 'chronic crisis stage' and a 'crisis resolution stage'. In this second example, crisis management commences and ends similarly to that proposed by Richardson. Where it diverges however is through differentiating between acute and chronic effects of a crisis. Acute effects onset suddenly following some incident which disrupt operations and threaten damage, life or limb. These must be supressed urgently if the crisis is to be resolved. Chronic effects, by contrast, linger in the wake of the acute effects. Although less dramatic, they can be just as disruptive, so also require suppression before the crisis may be considered to have been fully resolved.

The example presented in the fourth column is the four stage, two-layered matrix variant proposed by William Crandall, John A. Parnell and John E. Spillan.[89] It comprises a 'landscape survey stage', a 'strategic planning stage', a 'crisis management stage' and an 'organisational learning stage'. Each stage addresses not only the internal landscape, but the external landscape as well. In this third example, crisis management commences with proactive evaluations by respondents of internal and external threats against relevant crisis objects. An internal planning process then follows, with the crisis respondents developing strategies for dealing with any detected threats. This includes factoring in the external plans already made by other organisations such as industry regulators and emergency services. When a crisis eventually does strike, the next stage involves the respondents resolving disruption through engaging primary and secondary stakeholders and resources from both the internal and external landscapes. Once the crisis is resolved, the final stage entails a process of reflexive learning by the respondents about what transpired to improve the internal response. Again, this includes activities undertaken by supporting external organisations.

Lastly, the example presented in the fifth column is the five-stage, linear variant proposed by Christine M. Pearson and Ian I. Mitroff.[90] It comprises a 'signal detection stage', a 'preparation/prevention stage', a 'containment and damage control stage', a 'business recovery stage' and a 'learning stage'. In this fourth and final example presented here, crisis management commences with respondents undertaking proactive searches for crisis signals relating to relevant crisis objects. Once crisis signals have been detected, the respondents act on them, initially to prevent a crisis from emerging. Again, on the occasion that prevention proves impossible, however, the crisis respondents pivot to mitigation. In this scenario, the crisis respondents must next contain and limit the effects. Containment prevents the effects from spreading, while limitation minimises the extent that the effects can cause damage. Once the effects have been dealt with, the crisis respondents turn their attention to restoring operations. And finally, once the crisis has been resolved, the respondents undertake a process of learning with the aim of assessing the strengths and weaknesses of the response.

Each of these frameworks elucidate the processes that comprise crisis management. As with the many other crisis management frameworks that exist, they possess their own strengths and weaknesses of clarity, comprehensiveness and prescriptiveness in their recommendations for tackling crises. The crisis management framework deemed most useful for this study however is the fourth example proposed by Pearson and Mitroff. This manages to achieve a good balance between insight and practicable application, offering breadth and depth as well as flexibility. Neither is the guidance overly limited or prescriptive. Overall, it makes for a useful tool in analysing and understanding the process of crisis management.

Resilience

A consideration of the concepts of crisis and crisis management necessitates a consideration of the concept of 'resilience', another principle driving concept used in this study. Although rarely qualifiable until after the fact, resilience concerns the resources and processes by which some museum or community successfully traverses adverse environmental conditions.[91] There are three dominant understandings of resilience in this regard, identified by Ron Martin and Peter Sunley. One is the 'bounce back' understanding, where a system 'rebounds' to a 'pre-shock state or path'. Another is the 'ability to absorb' understanding, which '[e]mphasises [the] stability of system structure, function and identity in the face of shocks'. And the third is the 'positive adaptability' understanding, where a system 'maintain[s]' core performances despite shocks by adapting its structure'.[92] When these understandings are synergised and considered in the organisational context, resilience can be understood as the ability to restore from, withstand or adapt to difficult or changing conditions while continuing to maintain operationality, as required.[93] More fundamentally, it concerns the qualities of organisational flexibility, nimbleness and adaptability.[94] A resilient museum or community, therefore, is one which can diverge and diversify depending on societal need.

It is easy to confuse activities undertaken towards upholding a museum or community's resilience with activities undertaken towards resolving a crisis. After all, fundamentally, any museum or community which has successfully managed crisis will have done so by drawing on their resilient attributes. This assertion is supported by the work of Erica Seville, who reveals that the ingredients necessary for organisational resilience are also those essential to carrying out museum crisis management: 'situational awareness', 'proactive posture', 'planning strategies', 'decision making', *etcetera*.[95] But not all museums or communities which have found themselves undertaking such crisis management-related activities will have found themselves actually locked in the throes of crisis.

Although the acts of drawing on resilience and undertaking crisis management are both essentially reactional processes, the process performed in the context of resilience is routine, whereas the process performed in the context of crisis occurs under extraordinary circumstances. As discussed above, crises are uncomfortable, unpredictable, unstable and potentially dangerous situations which can strike any museum. Hence, while crisis management always involves addressing some sort of turbulence, confronting turbulence alone does not necessarily mean that a museum or community is gripped by crisis. For this, the turbulence must be disruptive, requiring extraordinary intervention to be overcome. It follows that resilience and crisis management are two separate though not mutually exclusive ideas. The former is the quality of being able to respond to change in the environment; the latter is the act of responding to critically disruptive events resulting from societal change.

Reinvention

A Product of Misalignment with the Social System

The third and final principle driving concept used in this study is 'reinvention'. It concerns the act of producing something new from an already existing entity. Reinvention is framed herein as a strategy for crisis management. Specifically, it comprises an approach to the management of revolutionary crises discussed above. This is where a crisis becomes understood through a modernity-esque lens as a breaking point and transitional phase.

The word reinvention has been variously defined despite its more recent proliferation in the English corpus compared with the word crisis. All published definitions however are firmly grounded in the concept of 'change': the transition from one state to another.[96] Moreover, mainstream English language dictionary definitions for the root word, 'reinvent', provide near uniform meanings: a process whereby a wholly new or radically different entity is conceived and created from something which already exists. These definitions are accurate insofar as they sketch out the idea. As with the word crisis however, each are too simplistic for use in academic work. Realistically, reinvention is a complicated and multifaceted process. Illustrating this point, work by Anthony Elliot demonstrates how both tangible and intangible man-made entities can be reinvented, while Montgomery Van Wart's work emphasises that individual processes are unique to the entity being reinvented at any given time or and within any given space.[97] Accordingly, because it involves both physical and non-physical entities, and also because no two processes are alike, reinvention requires a more sophisticated appreciation than one predicated on change alone.

Reinvention stems from a need for a utilitarian entity to change or do something completely differently. There could be many reasons for this requirement. When examining individual instances of reinvention however, the need derives from the fact that the entity in question, such as a museum, no longer meets expectations, meaning it sustains declining relevance. To be relevant is to yield positive cognitive effect: to make a positive difference to an individual's lived experience.[98] Accordingly, when museums stop yielding positive cognitive effect, they decreasingly benefit the society which they inhabit and invariably start becoming a drain on its stakeholder/s, without offering anything productive in return.

Museums generally experience widespread diminishing positive cognitive effect after their values become misaligned with those held by the social system they inhabit. This takes place when the expectations of one diverge away from the other. If caused by the museum, such divergences will most likely be resultant from organisational sclerosis or poorly conceived strategy. But if down to society, they are a natural and expected phenomenon. This is because societies embody fluid and continually transforming social systems. They develop new values through adopting fresh ideas and technologies that alter the way people act and interact within them. Events and trends can also force societies to modify their thoughts and practices. Van Wart calls this superordinate change a 'value shift'.[99] Depending on circumstances, some societies can undergo more frequent value shifts than others.[100]

When divergences occur between museums and society, a process of strategic realignment must take place if the museum is to restore its positive cognitive effect. Usually (though not necessarily) this involves the museum undergoing some adaption. Van Wart calls this counterweighting subordinate change a 'value adjustment'.[101] Through making value adjustments, institutions and organisations can adapt themselves to meet the updated needs held by society. But value adjustments rarely occur in parallel with value shifts. Various factors often discourage non-urgent adaption from taking place. Typically, value adjustments only occur when the pressure to adapt becomes so great that continuing with the *status quo* becomes impossible. For this reason, museums which do not make the requisite value adjustments when it is necessary ultimately risk placing their future in jeopardy.[102]

A museum's ability to make value adjustments also occurs through another sociological phenomenon called 'reflexivity'. This concept concerns the act of self-referring. It is a process by which actors, either acting for themselves or another entity, receive and react to relevant incoming information.[103] Depending on their interpretation of that information, the actors will then take whatever action is deemed appropriate in navigating the short-, medium- and long-term future. Emphasising this behavioural effect, Anthony Giddens calls the process 'reflexive monitoring of action'.[104] Reflexivity is a crucial ingredient of reinvention, indeed crisis consciousness and crisis management. Without the ability to evaluate situations against perceptions of present or future contexts, targeted actions for desired change would be impossible. This means that similarly to crisis consciousness and crisis management, the process of reinvention relies on a process of subjective assessment.

Another dimension to reinvention is its continuity. Josh Linkner argues that 'reinvention isn't an event; it's a lifelong process'.[105] This is born from the unremitting nature of social and environmental change. As societies continually undergo value shifts in some form or another, reinvention will remain forever a looming imperative over most museums while they seek to preserve their relevance and long-term prospects.

Analysing Reinvention

Every instance of reinvention is manifested through a highly individual process. They arise from circumstances, follow courses and result in outcomes that are dependent on the context wherein they occur. For example, a revolutionary crisis that prompts one crisis object's reinvention may have different underlying causes to the revolutionary crisis which prompts the reinvention of another. Each reinvention process therefore must satisfy whatever aim has instigated them. This idiosyncrasy can make analysing reinvention very challenging. Yet there are common components to them that can be identified and analysed via reinvention formulae which aid the understanding of the process: its origins, nature, effect, *etcetera*.

The original (and most theoretical) reinvention formula was conceived by David Gleicher at the management consultancy Arthur D. Little during the 1960s. Profiled by Richard Beckhard in the 1975 winter issue of *Sloan Management Review*, this approach to critiquing reinvention has since been adopted and developed by numerous theorists.[106] The original formula comprised an equation, representing the factors that reinvention necessitates:

$$C = (ABD) > X$$

In this equation, the value 'C' denotes change, 'A' denotes the level of dissatisfaction with the prevailing situation, 'B' denotes an understanding of the desired state, 'D' denotes an awareness of the practical first steps necessary to reach the desired state, and finally 'X' denotes the associated cost. The equation shows that for reinvention to take place, three conditions must be met. The first is a catalysing level of dissatisfaction with the current situation. The second is a conception of the desired situation.

And the third is awareness of the practical first steps necessary to reach that desired or required situation. Once these values have been established and weighed up, the need for change must be greater than the cost incurred executing it. This cost is more than finance and can cover any resource.[107]

$$Change = \left(Dissatisfaction, Desired\ State\ and\ Practical\ First\ Steps\right) > Cost$$

Three points arise when reflecting on the implications of this formula in the organisational context. The first is that without dissatisfaction, such as the dissatisfaction which arises from a revolutionary crisis, change becomes an exercise undertaken purely for its own sake. This occurrence poses risks for organisations. It potentialises subsequent unnecessary dissatisfaction and, in doing so, further change to address that. But such occurrences are conceivably rare. Even change that takes place during times of stability typically arises from dissatisfaction with the *status quo*.[108] The second point is that where only certain organisational agents perceive the need for change, their first steps should be to enrol those who do not see the need for change into empathising with and supporting these desires. After all, a united, supportive, consensual approach towards change stands far greater chance of succeeding than one which has partial or discordant support.[109] And the third point is that any desired organisational state must ultimately cohere with the values kept and upheld by the prevailing social system. If not, the change will unlikely be legitimised by stakeholders or society at large, potentially jeopardising the organisation through inactivity.[110]

Conclusion

The current chapter has explored and critiqued eight concepts and various complementary issues and ideas that constitute the analytical framework underpinning this study. First to be considered was institutions, organisations and museums. This established the conceptual context of the study, given how the study's case, the Imperial War Museum, embodies each construct. As organisations, museums embody institutions with the power to shape and move society. And yet, through comprising organisations, museums are not endowed with the same power as intangible institutions. This is because organisations must ensure they incorporate the ideas and practices of intangible institutions on the authority of people from society, which those people then look for in the social system. Hence, they are susceptible to societal whim—the context for certain crises and institutional reinvention.

In the second through fifth sections, the concepts of legitimacy, crisis and crisis management, resilience and reinvention were considered, alongside other relevant issues and ideas during each section. This set out the philosophy, assumptions and positionalities underlying the study as well as the interconnectedness and contributions of each concept used herein *apropos* the ways crises impact museums and museums resolve crises. It also contextualised the three foremost conceptual lenses applied against the case.

The concept of crisis is complex and challenging to comprehend and handle. When incorporated in research, scholars must study it thoroughly. This includes the word's etymology and different readings of the concept that have arisen over the years. Notwithstanding being an inherently subjective phenomena, crisis-conducive situations comprise difficult, fundamentally disruptive and potentially dangerous situations. When managing legitimised crises, museums and their communities must draw on attributes of resilience to overcome adversity. These can be present across their physical and non-physical elements and should be sufficiently developed so that they can withstand diverse and harsh conditions. Occasionally however, drawing on existing resilient attributes is not sufficient to overcome a crisis, especially in the case of institutions and organisations. Sometimes, the only action which will suffice involves the institution or organisation completely reinventing.

Reinvention is a younger word than crisis. Even so, the concept, which embodies ideas and principles that far predate the word's creation and proliferation in the English corpus, requires no less an in-depth analysis to be fully appreciated. Fundamentally, however, reinvention concerns the process of creating something profoundly new from something already in existence. For institutions and organisations that face a crisis of diminishing relevance and legitimacy, reinvention can be an effective approach to managing it. Such situations manifest from diminished coherence between an institution or organisation and the social system they inhabit. Analysing this process however poses a challenge to researchers because of reinvention's variability. Fortunately, they can overcome the problem by investigating the issues and ideas driving it: dissatisfaction with the *status quo*, a contrasting desired state and the practical steps necessary towards reaching that desired state.

The current chapter has presented thinking behind this study on the Imperial War Museum during the Second World War era. The following six chapters present the study itself. In the next chapter, the Imperial War Museum is historically and conceptually contextualised as an institution, organisation and museum during the years between the two world wars. Drawing on relevant issues and ideas discussed above, it serves to show that the Imperial War Museum of the 1930s and 1940s embodied many current institutional, organisational and museological features, demonstrating the contemporary relevance this study has for both professionals and scholars concerned for museums and their longevity.

Notes

1. Rhiannon Mason, ed., *Museum Studies*, 5 vols (Abingdon: Routledge, 2019); Christer Pursiainen, *The Crisis Management Cycle* (Abingdon: Routledge, 2018).
2. David A. Snow, E. Burke Rochford Jr, Steven K. Worden and Robert D. Benford, 'Frame Alignment Processes, Micromobilization, and Movement Participation', *American Sociological Review* 51 (1986): 469.
3. Ibid.
4. Ibid., 467.
5. Joseph Henri Jupille and James A. Caporaso, *Theories of Institutions* (Cambridge: Cambridge University Press, 2022), 2.
6. Ibid., 3–6.
7. W. Richard Scott, John W. Meyer and associates, *Institutional Environments and Organizations: Structural Complexity and Individualism* (Thousand Oakes, CA: SAGE, 1994), 68.
8. Tim Hallett and Marc J. Ventresca, 'Inhabited Institutions: Social Interactions and Organizational Forms in Gouldner's "Patterns of Industrial Bureaucracy"', *Theory and Society* 35 (2006): 215.
9. Peter Berger and Thomas Luckmann, *The Construction of Reality: A Treatise in the Sociology of Knowledge* (London: Penguin, 1991), 93.
10. Allen Buchanan, 'Institutional Legitimacy', in *Oxford Studies in Political Philosophy*, vol. 4, ed David Sobel, Peter Vallentyne and Steven Wall (Oxford: Oxford University Press, 2018), 53–79.
11. William H. Sewell Jr, 'A Theory of Structure: Duality, Agency, and Transformation', *American Journal of Sociology* 98 (1992): 1–29.
12. Paul Griseri, *An Introduction to the Philosophy of Management* (London: SAGE, 2013), 9–22.

13. James March, Erhard Friedberg and David Arellano, 'Institutions and Organizations: Differences and Linkages from Organization Theory', *Gestión y Política Pública* 20 (2011): 235–246.
14. Elias L. Khalil, 'Organizations Verses Institutions', *Journal of Institutional and Theoretical Economics* 151 (1995): 499.
15. David A. Buchanan and Andrzej A. Huczynski, *Organizational Behaviour*, 9th edn (Harlow: Pearson Education, 2017), 8–9.
16. Gareth R. Jones, *Organizational Theory: Text and Cases*, 3rd ill. edn (Upper Saddle River, NJ: Pretence-Hall, 2001), 2.
17. Stephen Sheppard, 'Museums in the Neighborhood: The Local Economic Impact of Museums', in *Handbook of Industry Studies and Economic Geography*, ed Frank Giarratani, Geoffrey J. D. Hewings and Philip McCann (Cheltenham: Edward Elgar, 2013), 191–204.
18. Ntak George Akpanama and Chukwuemeka Ekenta, 'Examining the Negative Impact of Abattoir Operations on Residential Property Value within Abattoir Site in Port Harcourt Metropolis', *International Journal of Innovative Science, Engineering and Technology* 9 (2022): 26–34.
19. W. Richard Scott, *Institutions and Organizations: Ideas, Interests and Identities*, 4th edn (Thousand Oakes, CA: SAGE, 2014), 182–183.
20. Douglas North, *Institutions, Institutional Change and Economic Performance* (Cambridge: Cambridge University Press, 1990).
21. Philip Selznick, *Leadership in Administration: A Sociological Interpretation* (Berkeley, CA: University of California Press, 1984); Oliver E. Williamson, *Markets and Hierarchies—Analysis and Antitrust Implications: A Study in the Economics of Internal Organization* (New York, NY: Free, 1975); Oliver E. Williamson, *The Economic Institutions of Capitalism: Firms, Markets, Relational Contracting* (New York, NY: Free, 1985); Oliver E. Williamson, 'Transaction Cost Economics and Organization Theory', *Industrial and Corporate Change* 2 (1993): 107–156.
22. Staffan Furusten, *Institutional Theory and Organisational Change* (Cheltenham: Edward Elgar, 2013), 65–78.
23. John W. Meyer and Brian Rowan, 'Institutionalized Organizations: Formal Structure as Myth and Ceremony', *American Journal of Sociology* 83 (1977): 340–363; Lynne G. Zucker, 'Organizations as Institutions', in *Research in the Sociology of Organizations*, vol. 2, ed Samuel B. Bacharach (Greenwich, CT: JAI, 1983), 1–47; Frank R. Dobbin, 'Cultural Models of Organisation: The Social Construction of Rational', in *The Sociology of Culture: Emerging Theoretical Perspectives*, ed Diana Crane (Cambridge, MA: Blackwell, 1994), 117–141.
24. John Child, *Organization: Contemporary Principles and Practice* (Malden, MA: Blackwell, 2005), 399.
25. Jupille and Caporaso, *Theories of Institutions*, 7.

26. John E. Simmons, *The Museum: A History* (Lanham, MD: Rowman and Littlefield, 2016).
27. Rhiannon Mason, Alistair Robinson and Emma Coffield, *Museum and Gallery Studies: The Basics* (Abingdon: Routledge, 2018), 39–41.
28. Beatriz Plaza and Silke N. Haarich, 'Museums for Urban Regeneration? Exploring Conditions for Their Effectiveness', *Journal of Urban Regeneration and Renewal* 2 (2009): 259–271; Fiona Tuck and Scott Dickinson, *The Economic Impact of Museums in England* (Newcastle upon Tyne: TBR, 2015), https://www.artscouncil.org.uk/sites/default/files/download-file/Economic_Impact_of_Museums_in_England_report.pdf (18 March 2024).
29. Steven Cooke and Lloyd Jenkins, 'Discourses of Regeneration in Early Twentieth-Century Britain: From Bedlam to the Imperial War Museum', *Area* 33 (2001): 382–390.
30. For example, Fiona Candlin, 'Defining Museums', Blog, Mapping Museums Lab, https://mapping-museums.bbk.ac.uk/2017/10/30/defining-museums/ (16 August 2024).
31. Eilean Hooper-Greenhill, *Museums and the Shaping of Knowledge* (London: Routledge, 1992), 1.
32. Timothy Ambrose and Crispin Paine, *Museum Basics: The International Handbook*, 4th edn (Routledge: Abingdon, 2018), 9–11.
33. John Fraser, 'A Discomforting Definition of Museum', *Curator: The Museum Journal* 62 (2019): 501–504.
34. David Lowenthal, 'Patrons, Populists, Apologists: Crises in Museum Stewardship', in *Valuing Historic Environments*, ed Lisanne Gibson and John Pendlebury (Farnham: Ashgate, 2009), 19–31.
35. Kenneth Hudson, 'The Museum Refuses to Stand Still', *Museum International* 50 (1998): 48.
36. Graham Black, 'Introduction: The Challenge of Change', in *Museums and the Challenge of Change: Old Institutions in a New World*, ed Graham Black (Abingdon: Routledge, 2021).
37. Simon J. Knell, Suzanne Macleod and Sheila Watson, 'Introduction', in *Museum Revolutions: How Museums Change and Are Changed*, ed Simon J. Knell, Suzanne Macleod and Sheila Watson (Abingdon: Routledge, 2007), xix–xxvi.
38. Roy Suddaby, Alex Bitektine and Patrick Haack, 'Legitimacy', *Academy of Management Annals* 11 (2017): 451–478.
39. Mark C. Suchman, 'Managing Legitimacy: Strategic and Institutional Approaches', *Academy of Management Review* 20 (1995): 574.
40. Clive Gray, *The Politics of Museums* (Basingstoke: Palgrave Macmillan, 2015), 18–23.

41. Carol A. Scott, ed., *Museums and Public Value: Creating Sustainable Futures* (Abingdon: Routledge, 2016).
42. Tony Bennett, *The Birth of the Museum: History, Theory, Politics* (London: Routledge, 1995), 26–27; Catherine Pearson, *Museums in the Second World War: Curators, Culture and Change*, ed Suzanne Keene (Abingdon: Routledge, 2017), 25–26; Clive Gray, 'Instrumental Policies: Causes, Consequences, Museums and Galleries', *Cultural Trends* 17 (2008): 209–222.
43. Marcelo Bucheli and Jin Uk Kim, 'The State as a Historical Construct in Organizational Studies', in *Organizations in Time: History, Theory, Methods*, ed Marcelo Bucheli and R. Daniel Wadhwani, pbk edn (Oxford: Oxford University Press, 2015), 252.
44. Jeffrey Abt, *Valuing Detroit's Art Museum: A History of Fiscal Abandonment and Rescue* (Cham: Palgrave Macmillan, 2017).
45. Graham Matthews, Yvonne Smith and Gemma Knowles, *Disaster Management in Archives, Libraries and Museums* (Aldershot: Ashgate, 2009); Daniel J. Benny, *Cultural Property Security: Protecting Museums, Historic Sites, Archives, and Libraries* (Boca Raton, FL: CRC, 2013); Brent A. Powell, *Collection Care: An Illustrated Handbook for the Care and Handling of Cultural Objects* (Lanham, MD: Rowman and Littlefield, 2016).
46. Roger O'Keefe, *The Protection of Cultural Property in Armed Conflict* (Cambridge: Cambridge University Press, 2006).
47. Suddaby, Bitektine and Haack, 'Legitimacy', 451.
48. Rüdiger Graf and Konrad H. Jarausch, '"Crisis" in Contemporary History and Historiography', *Docupedia-Zeitgeschichte* (2017): para. 1, 10.14765/zzf.dok.2.789.v1, https://zeitgeschichte-digital.de/doks/frontdoor/deliver/index/docId/789/file/docupedia_graf_jarausch_crisis_v1_en_2017.pdf (18 March 2024).
49. Reinhart Koselleck, 'Crisis', trans. Michaela W. Richter, *Journal of the History of Ideas* 67 (2006): 358–361, 397–400; Andrea Borghini, 'A Reassessment of the Category of Crisis: Some Reflections from the Social Sciences', in *Exploring the Crisis: Theoretical Perspectives and Empirical Investigations*, ed Andrea Borghini and Enrico Campo (Pisa: Pisa University Press, 2015), 325–326.
50. Graf and Jarausch, '"Crisis"', para. 3; Florian Ravail, 'The Crisis: Meaningless Catch-all Term or Rigorous Concept?', trans. Thomas Williams, *Kinea* (2018), para 2 (15 April 2018), note that this reference is no longer available as the online publication has been discontinued.
51. Graf and Jarausch, '"Crisis"', para. 1.
52. Borghini, 'Reassessment', 342.

53. Brian Milstein, 'Thinking Politically About Crisis: A Pragmatist Perspective', *European Journal of Political Theory*, 14 (2015): 142.
54. Stuart Hall, 'The Work of Representation', in *Representation*, ed Stuart Hall, Jessica Evans and Sean Nixon, 2nd edn (London: SAGE, 2013), 7–10.
55. Koselleck, 'Crisis', 358–360.
56. Ibid., 360–361.
57. Milstein, 'Thinking Politically', 145; see Thomas Hobbes, *Leviathan; or, The Matter, Forme and Power of a Common Wealth Ecclesiasticall and Civil* (London: Andrew Crooke, 1651).
58. Milstein, 'Thinking Politically', 145.
59. Thomas Paine, *The American Crisis: Number 1* ([Philadelphia?]: [Styner and Cist?], 1776), para. 13.
60. Angelos Mouzakitis, 'Modernity and the Idea of Progress', *Frontiers in Sociology* 2 (2017): 10.3389/fsoc.2017.00003, https://www.frontiersin.org/articles/10.3389/fsoc.2017.00003/full (18 March 2024).
61. Elliott Johnson, David Walker and Daniel Gray, *Historical Dictionary of Marxism*, 2nd edn (Lanham, MD: Rowman and Littlefield, 2014), s.v. 'crises'.
62. Christopher Whitehead, Mads Daugbjerg, Susannah Eckersley and Gönül Bozoğlu, 'Dimensions of European Heritage and Memory: A Framework Introduction', in *Dimensions of Heritage and Memory: Multiple Europes and the Politics of Crisis*, ed Christopher Whitehead, Susannah Eckersley, Mads Daugbjerg and Gönül Bozoğlu (Abingdon: Routledge, 2019), 2.
63. Milstein, 'Thinking Politically', 147.
64. Ibid., 147–152.
65. Mary Douglas, *Purity and Danger: An Analysis of Concepts of Pollution and Taboo* (New York, NY: Frederick A. Praeger, 1966), 35.
66. Vener Garayev, 'Crisis, Definition of', in *Encyclopedia of Crisis Management*, ed K. Bradley Penuel, Matt Statler and Ryan Hagen (Thousand Oaks, CA: SAGE, 2013), 186–187.
67. Shari R. Veil, 'Routine Emergencies versus True Crises', in *Encyclopedia of Crisis Management*, ed K. Bradley Penuel, Matt Statler and Ryan Hagen (Thousand Oaks, CA: SAGE, 2013), 845.
68. Michele Wucker, *The Gray Rhino: How to Recognize and Act on the Obvious Dangers We Ignore* (New York, NY: St Martin's, 2016).
69. Ravail, 'The Crisis', paras 21–23.
70. Garayev, 'Crisis, Definition of', 186–187.
71. Nassim Nicholas Taleb, *The Black Swan: The Impact of the Highly Improbable* (London: Allen Lane, 2007).
72. Didier Sornette, 'Dragon-Kings, Black Swans and the Prediction of Crises', *International Journal of Terraspace Science and Engineering* 2

(2009): para. 21–23, 10.48550/arXiv.0907.4290, https://arxiv.org/ftp/arxiv/papers/0907/0907.4290.pdf (18 March 2024).
73. Veil, 'Routine Emergencies', 845; Ziqiang Han and Qiujie Zhang, 'Disruption of Organizations', in *Encyclopedia of Crisis Management*, ed K. Bradley Penuel, Matt Statler and Ryan Hagen (Thousand Oaks, CA: SAGE, 2013), 292–294.
74. Graf and Jarausch, '"Crisis"', para. 39.
75. Ibid.
76. R. J. Overy, *The Inter-war Crisis, 1919–1939* (Harlow: Pearson, 1994), 1.
77. Graf and Jarausch, '"Crisis"', para. 12.
78. Max Kölbel, 'Faultless Disagreement', *Proceedings of the Aristotelian Society* 104 (2004): 53–73.
79. Graf and Jarausch, '"Crisis"', para. 38.
80. Joanne E. Hale, David P. Hale and Ronald E. Dulek, 'Decision Processes during Crisis Response: An Exploratory Investigation', *Journal of Managerial Issues* 18 (2006): 316.
81. Finn Frandsen and Winni Johansen, *Organizational Crisis Communication: A Multivocal Approach* (London: SAGE, 2017), 32; Ross Campbell, *Crisis Control: Preventing and Managing Corporate Crises* (Sydney, NSW: Prentice Hall, 1999), 23.
82. William 'Rick' Crandall, John A. Parnell and John E. Spillan, *Crisis Management: Leading in the New Strategy Landscape*, self-pub. 4th edn (n.place: n.pub., 2021), 9–18, note that this monograph was formally published by SAGE in 2014.
83. Jonathan Bernstein, *Manager's Guide to Crisis Management* (New York, NY: McGraw-Hill, 2011), 20–24.
84. Crandall, Parnell and Spillan, *Crisis Management*, 4.
85. Pursiainen, *Crisis Management Cycle*.
86. Bill Richardson, 'Socio-technical Disasters: Profile and Prevalence', *Disaster Prevention and Management* 3 (1994): 41–69.
87. Crandall, Parnell and Spillan, *Crisis Management*, 11.
88. Stephen Fink, *Crisis Management: Planning for the Inevitable* (New York, NY: American Management Association, 1986).
89. Crandall, Parnell and Spillan, *Crisis Management*, 13–18.
90. Christine M. Pearson and Ian I. Mitroff, 'From Crisis Prone to Crisis Prepared: A Framework for Crisis Management', *The Executive* 7 (1993): 48–59.
91. Ran S. Bhamra, Kevin Burnard and Samir S. Dani, 'Resilience: A Concept, a Literature Review and Future Directions', in *Organisational Resilience: Concepts, Integration and Practice*, ed Ran Bhamra (Boca Raton, FL:

CRC, 2015), 4; Diane L Coutu, 'How Resilience Works', *Harvard Business Review* 80 (2002): 47; Carry Cooper, Jill Flint-Taylor and Michael Pearn, *Building Resilience for Success: A Resource for Managers and Organisations* (Basingstoke: Palgrave Macmillan, 2013), 14–15.

92. Ron Martin and Peter Sunley, 'On the Notion of Regional Economic Resilience: Conceptualization and Explanation', *Journal of Economic Geography* 15 (2015): 4.
93. James J. Leflar and Marc H. Siegel, *Organizational Resilience: Managing the Risks of Disruptive Events – A Practitioner's Guide* (Boca Raton, FL: CRC, 2013), 11; Bhamra, Burnard and Dani, 'Resilience', 21; Ian P. McCarthy, Mark Collard and Michael Johnson, 'Adaptive Organizational Resilience: An Evolutionary Perspective', *Current Opinion in Environmental Sustainability* 28 (2017): 33.
94. Robert R. Janes, *Museums in a Troubled World: Renewal, Irrelevance or Collapse?* (Abingdon: Routledge, 2009), 141.
95. Erica Seville, *Resilient Organizations: How to Survive, Thrive and Create Opportunities through Crisis and Change* (London: Kogan Page, 2017), 19.
96. Liz Clarke, *The Essence of Change* (New York, NY: Prentice Hall, 1994).
97. Anthony Elliot, *Reinvention* (Abingdon: Routledge, 2013); Montgomery Van Wart, 'The First Step in the Reinvention Process: Assessment', *Public Administration Review* 55 (1995): 429–438.
98. Deirdre Wilson and Dan Sperber, 'Relevance Theory', in *The Handbook of Pragmatics*, ed Laurence R. Horn and Gregory L. Ward, 2nd edn (Maldon, MA: Blackwell, 2006), 608.
99. Van Wart, 'First Step', 429.
100. Ibid.
101. Ibid., 430.
102. Ibid.
103. Allan G Johnson, *The Blackwell Dictionary of Sociology: A User's Guide to Sociological Language*, 2nd edn (Maldon, MA: Blackwell, 2000), s.v. 'reflexivity'.
104. Anthony Giddens, *The Consequences of Modernity* (Cambridge: Polity, 1990), 36.
105. Josh Linkner, *The Road to Reinvention: How to Drive Disruption and Accelerate Transformation* (San Francisco, CA: Jossey-Bass, 2014), 11.
106. Steven H. Cady, Robert 'Jake' Jacobs, Ron Koller and John Spalding, 'The Change Formula: Myth, Legend, or Lore?', *OD Practitioner* 46 (2014): 32–39.
107. Richard Beckhard, 'Strategies for Large System Change', *Sloan Management Review* 16 (1975): 45.

108. Freek Vermeulen, Phanish Puranam and Ranjay Gulati, 'Change for Change's Sake', *Harvard Business Review* 88 (2010): 70–76.
109. Tim Hannagan, *Mastering Strategic Management* (Basingstoke: Palgrave Macmillan, 2002), 154–157.
110. John Dowling and Jeffrey Pfeffer, 'Organizational Legitimacy: Social Values and Organizational Behaviour', *Pacific Sociological Review* 18 (1975): 122–136.

CHAPTER 3

The Imperial War Museum, 1917–1939

Introduction

The issues and ideas presented in Chap. 2 indicate that crisis and change are two likelihoods which museums must contend with over their existences. They can occur at any time: an unstable social system increases their probability.[1] While the nature, extent and impact of a crisis cannot be known until after the fact, it is not unreasonable to assume that their outcomes may potentialise far reaching institutional, organisational and museological consequences.[2] The crisis-conducive situations experienced at the Imperial War Museum over the Second World War era, considered in the coming chapters, demonstrate this.

To fully appreciate what occurred at the Imperial War Museum during the Second World War, an awareness of its historical and conceptual development during the inter-war years is essential. After all, museums are more than just *museums*. They are also institutions and organisations—concepts and constructs which have philosophical and real-world implications for the way museums exist, work, and how crises impact them—and thus, to use Mara Cerquetti and Eleonora Cutrini's description, are 'a complex socio-technical system'.[3] These contexts are the focus of the current chapter. In exploring them, it characterises the Imperial War Museum as an institution, organisation and museum. Through doing so, it also gives profile to other key aspects of the case that feature prominently over later chapters.

P. W. Deans, *Crisis, Reinvention and Resilience in Museums*,
https://doi.org/10.1007/978-3-031-67806-6_3

The current, first case chapter addresses the above over three substantive sections. The first section considers the Imperial War Museum as an institution in the conceptual sense. It analyses the shifting sources of power and legitimacy of the museum before the Second World War, discussing the museum's dependence on society and society's influence on the museum. The second section then considers the Imperial War Museum as an organisation. It discusses the organisational architecture embodying the museum before the war: the museum's structure, processes and boundaries. And the third section before the chapter's conclusion considers the Imperial War Museum as a museum. It discusses the collections and exhibitions of the museum—arguably the most indicative and distinctive elements to any museum—before the war, analysing their basic features, traits and qualities.

The Imperial War Museum as an Institution

Three Pillars of Institutionalisation

Institutions in the conceptual sense can be understood as physical and abstract constructs that guide, influence or control how society operates and generates knowledge and understanding about itself. They also advance and perpetuate societal behaviour.[4] In doing so, institutions bring structure, order and stability to society by establishing the 'rules of the game' that direct social life and by fostering its 'nature of play'. This is achieved through incorporating what W. Richard Scott conceptualises as 'regulative', 'normative' and 'cultural-cognitive' institutional elements: the 'pillars' which metaphorically uphold institutions.[5] Each pillar establishes a basis for institutional power. They also determine whether power is exercised openly or secretly and implemented forcefully, coercively or consensually.[6]

It is not uncommon for institutions to rest on two or three pillars at any given moment, for one pillar to be perceived as dominant, or for different people to perceive dominance in different pillars. This stems from the fact that when institutional functionality changes, so too can conceptions about the predominance of different pillars, and the fact that individual conception regarding any such predominance depends on the beholder's personal ontological and epistemological positionality.[7] Most noteworthy, it is also not uncommon for institutions to (be seen to) change their power base. As the social system experiences value shifts, an institution supported by one pillar may eventually find itself sustained by another.[8]

A Regulative Institution, 1917–1918

On 5 March 1917, at the height of the First World War, the War Cabinet approved a proposal by government minister Sir Alfred Mond, the First Commissioner of Works, to establish a state-sponsored war museum in the United Kingdom.[9] Formally conceived as the National War Museum, there are two arguments for why this decision was made before the conflict's conclusion. One is forwarded by Gaynor Kavanagh, a museologist and museum historian. Another is forwarded by Diana Condell, a former curator at the Imperial War Museum.

Kavanagh's argument stems from two political events that occurred during late 1916 and early 1917 in the United Kingdom. The first was a political coup on 6 December, which saw David Lloyd George replace Herbert Asquith as the Prime Minister. The second was Lloyd George's subsequent restructuring of the state's propaganda machine on 20 February.[10] After assuming office, the new Prime Minister wasted no time in consolidating his position. He had perceived the war effort to flounder from low morale caused by high casualties and increasing privations, and realised the importance opinion forming had in combatting this. He therefore revamped the country's prosecution of the war, overhauling various government departments and ministries.[11] Most relevant in this regard was his aforementioned work improving propaganda. After receiving reports about the disorganised operation at the existing propaganda bureau, Wellington House, Lloyd George arranged for the War Cabinet to approve its supersession. In place was established the Department of Information: a more focussed, less fragmented body, with greater scope for domestic opinion forming.[12] This occurred two weeks before the 5 March decision to found a national war museum. Kavanagh questions whether 'the National War Museum was part of a much larger initiative on propaganda, aimed at combatting war weariness'.[13] The establishment of the National War Aims Committee, five months after the 5 March decision, adds credence to this argument.[14]

Condell's argument, by contrast, stems from the cultural effects of the tremendous loss of life sustained by the British Army over the months leading to the 5 March decision, particularly from the Battle of the Somme some eight months earlier.[15] During this offensive lasting 1 July–18 November 1916, around 250,000 soldiers, predominantly volunteers recruited at the war's outset, were killed or injured irrecoverably. An estimated 182,000 were also temporarily incapacitated.[16] Compounding matters, the casualties often came from concentrated areas, the men having been recruited into so-called pals battalions at the war's outset.[17] Condell contends that tragedies like this, alongside other impacts which the war had on society, fundamentally altered 'the relationship between citizen and the state': 'for the first time, [the country fought] a truly national war'.[18] As a result, people started searching for conduits through which to commemorate the dead. When these could not be found, they began creating their own by erecting street shrines while local newspapers and other periodicals printed 'rolls of honour'.[19] Clearly, some state recognition was needed.[20]

These arguments presented by Kavanagh and Condell for the foundation of the National War Museum draw on different reasoning. Hence, their conceptualisations regarding the museum as an institution totally differ. Kavanagh argues that the National War Museum arose from political action to reinvigorate the country's fighting spirit. Accordingly, she implies the creation of an institution resting largely on the regulative pillar.[21] Regulative institutions can be understood as institutions which dictate, manage and promote societal conduct. They establish rules and codes for individuals and communities, monitor compliance, and issue reward or punishment for any adherence or breaches.[22] This conclusion contrasts with the one arising from Condell, who argues that the National War Museum was founded to satisfy a public need for some formal commemorative outlet. She, by contrast, implies the creation of an institution geared towards socio-cultural healing and cohesion: one resting largely on the cultural-cognitive column, which is discussed more below.[23]

It can be confidently speculated that both cultural and political need were factored in the final decision to establish the National War Museum. Yet, when weighing up the evidence, the more plausible stimulus comprised the political, suggesting the decision derived mainly, as Kavanagh argues, from operational requirement. Additional evidence supporting this view comes from the fact that during 1916, the government had closed all national museums and galleries for cost cutting purposes until after the war.[24] It had also tried unsuccessfully to halt the construction of the only art gallery built during the war years themselves, the Shipley Art Gallery in Gateshead.[25] Moreover, the initial sum agreed by the government for the project was set at just £3000.[26] Such limited expenditure implies an expectation of quick returns. By contrast, in 1823 the National Gallery received £60,000.[27] And finally, the project was initiated and, until 2 July 1920, funded and operated without formal incorporation. As such, when formal incorporation was eventually realised via an Act of Parliament, there remained no need to explicitly provide for the establishment amongst the legislation's text.[28] This is because the museum had already been established *de facto* via a Class IV vote in Parliament on 27 February 1918.[29] According to Kavanagh's interpretation of events, the move afforded it 'some recognition [...] as a formal, national institution'.[30] By contrast, the Act of Parliament that incorporated the National Maritime Museum 14 years later did, by necessity, provide for that institution's establishment, what with there being no pre-existing embryonic or appropriately comparable body which could be co-opted.[31]

All this evidence points to the National War Museum originally being conceived by the War Cabinet as a short-term regulative institution. Yet, clearly, it was not one geared towards setting and enforcing codified laws and penalties. Founded to reinvigorate the country's national war effort, the museum was rather an institution geared towards stimulating societal conduct. This occurred by deploying far softer mechanisms, such as targeting emotion. The institution therefore proceeded over the remaining war years by empowering action through inspiring people to act and by fostering guilt and shame amongst those who did not.[32] The initial wartime temporary exhibitions of the museum exemplify this.[33] For example, at Burlington House and Whitechapel Art Gallery, women's war work was heavily profiled with the view to recruiting women into the various women's wartime services. At Whitechapel Art Gallery particularly, live costumed interpretation made visitors feel much more like participants in the war, rather than spectators.[34]

A Normative Institution, 1918–c. 1933

The First World War and the original regulative imperative of the National War Museum ended on 11 November 1918. This led to diminishing public and government support for the project.[35] But the organising committee pressed on with their work, rather than wind it down for having become obsolete. One significant decision taken at this time involved replacing the word 'National' with 'Imperial' in the museum's title, an ostensibly inclusive move, made following recommendations by the Dominions subcommittee to broaden the initiative's appeal.[36] Thus, the Imperial War Museum opened on 9 June 1920 at the Crystal Palace, Sydenham, with a colonial purview and a purpose resembling a normative institution. Normative institutions uphold the aims and objectives of society through introducing rigid, mandatory dynamics which help society assess usefulness, importance and worth in the social system while setting the methodology necessary for obtaining them.[37] They hold similarities with regulative institutions by comprising entities which influence societal conduct.[38] The difference, however, lies in their efforts to normalise ideas and practices, rather than impose them.[39] Normative institutions achieve this through promoting and disseminating 'values' and 'norms'. Values are issues and ideas held by individuals, communities or organisations to be crucial or beneficial. Norms, in contrast, are the routine means by which values become acquired. Considering this, it *could* be argued that the Imperial War Museum was not predominantly a regulative institution during the First World War, but a normative one. That argument is not subscribed to here, however. This is because regulative institutions possess an instrumental logic, the apparent rationale of the Imperial War Museum until November 1918.[40] In comparison, normative institutions embody a logic geared towards setting out what is and is not appropriate for society: whether they cohere with accepted values and norms.[41]

One clear way that the Imperial War Museum comprised a normative institution immediately following the First World War occurred through nurturing discourses that honoured the British national and imperial war effort. The tone in this regard was set at its opening ceremony on 9 June 1920, attended by royalty, clergy, politicians and other important dignitaries. Reports describe an event underpinned with solemn Christian acts.[42] It was a symbolic moment, marking what many believed to be the start of an enlightened, more peaceful age. Sue Malvern similarly interprets the event as 'a funeral or laying-to-rest of the nation's trauma and loss'.[43] Addressing King George V at the opening, Sir Alfred Mond spoke how: 'The Museum was not conceived as a monument of military glory, but rather as a record of toil and sacrifice'.[44] In response, The King declared that:

> We cannot say with what eyes posterity will regard this Museum, nor what ideas it will arouse in their minds. We hope and pray that as the result of what we have done and suffered they may be able to look back upon war, its instruments and its organisation as belonging to a dead past.[45]

Such sentiments at the time came to epitomise the mission of the Imperial War Museum over the inter-war years: that of making a representation of the 'war to end all war'.

Proceeding in this vein throughout the inter-war years, the Imperial War Museum presented a message that the Allied victory over the Central Powers had been necessary but costly for the country, involving major losses of life and financial and material resources. Through doing so, it became a mnemonic space by curating exhibitions aiding memory and remembrance. The museum reminded visitors that the war had been hard-won through a united effort. It avoided triumphalism, but was fundamentally antagonistic in nature, justifying why the United Kingdom had become involved and celebrating the military achievements which led to the final outcome.[46] The nuanced interpretations of the exhibitions however complicated this message.[47] In one sense they were commemorative, fitting with the celebratory presentation.[48] The objects came to be treated as 'sacred relics':[49] a term from Christian theology for an object comprising the physical remains of, or which has been touched by, a saint.[50] But in another sense they were also technical. At the same time, the interpretation profiled scientific and engineering detail.[51] The Director-General's preamble to the 1920–1921 annual report describes the balancing act: 'The personal and historical value of certain of the exhibits is emphasised, but it is never allowed to detract or overshadow their value as technical illustrations set before the serious student'.[52] Alys Cundy rationalises this by conceptualising them as possessing an interpretive hybridity: 'The museum's objects were perceived to function as items of technical education and interest, but also as "actual" remains with a physical connection to the past'.[53] Overall, visitors faced manifold meanings.

The norms which the Imperial War Museum instituted to achieve these values involved instilling an air of remembrance. It must be noted that the museum was not, nor ever has been, an official memorial. Yet throughout the inter-war years, the Imperial War Museum did comprise a *de facto* centre for commemoration. This resulted not only from the commemorative qualities maintained by the exhibits discussed above, but also the practice at the museum of scattering memorialising items throughout the galleries and other spaces such as entrances, stairways, corridors and vestibules.[54] On occasion, both historical exhibits and memorialising items were exhibited together to impactful effect. Drawing on Michel Foucault's ideas about 'heterotopia'—the concept that certain defined space can hold meaning which juxtaposes, disturbs or intensifies the space surrounding it[55]—Cundy suggests such arrangements became used 'to establish a system of representation that differed from yet also transcended, and thus provided the backdrop to, the informative main displays'.[56]

In one early inter-war example of this unofficial approach to memorialisation at the Crystal Palace, people visiting the Imperial War Museum could see the original temporary wood and plaster cenotaph designed by Sir Edwin Landseer Lutyens, placed alongside a preserved field gun from L Battery, Royal Horse Artillery, deployed at the Affair of Néry on 1 September 1914.[57] The gun on display was the only one from L Battery which survived the skirmish, its crew each being awarded the Victoria Cross for their actions.[58] At least one conceivable effect of the memorial's placement alongside the gun was a link between valour and sacrifice. In another, later inter-war example, occurring when the museum occupied the Imperial Institute, South Kensington, from 1924 until 1936, the same temporary cenotaph ended up being placed alongside other commemorative elements by the museum's entrance.[59] When visitors passed through this heterotopic, memorialised 'installation'—Saadi Lahlou's term for a socially constructed control apparatus[60]—they were targeted with sober memorialising sentiments. These sentiments would then be carried through the galleries by the visitors, conforming behaviour to the museum's desired standards of thought, meaning-making and deference. And finally, in yet another, even later inter-war example, occurring when the museum was located at the former-Bethlem Royal Hospital building, Southwark, where it has resided since 1933, the temporary cenotaph became placed alongside artwork at the Picture Gallery. Here, the museum conceivably instigated or filliped a commemorative reaction to the paintings.[61] Whether deliberate or not, such curatorial decision making rendered the Imperial War Museum a veritable sanctum of memorialisation.

Another, more uncomfortable, way that the Imperial War Museum *may* have comprised a normative institution immediately following the First World War was by buttressing white dominance in the popular narrative of the British Empire during the conflict. Steven Cooke and Lloyd Jenkins suggest that the museum's name change in 1918 may actually have been undertaken to help the authorities 're-establish[…] the boundaries between Imperial power and Empire'.[62] This interesting if unsettling idea derives from the received wisdom on the poor early inter-war race relations which were taking hold across the country at the time and, crucially, the authorities' conciliation of the pernicious white racism underlying the hostility. For example, Black soldiers were excluded from the 1919 Victory Parade, thereby deliberately effacing their contribution from public record and memory.[63] While noting that the museum's inter-war exhibitions certainly infer a British national dominance, more research is required before this idea can be confidently adopted, as Cooke and Jenkins offer no substantive empirical evidence to support their suggestion.

A Cultural-Cognitive Institution, c. 1933–1939

Throughout the inter-war years, the Imperial War Museum echoed British society's concerns for peace and rejecting armed conflict. It did this by increasingly profiling anti-war messages about war's futility alongside the commemorative-technical messages present from the earliest years. In doing so, the museum became ever more reliant on the cultural-cognitive pillar to make relevant representations. Cultural-cognitive institutions are those which represent the cultural frameworks that make up the social system. They are built on and perpetuate 'orthodoxy': what society considers right and proper.[64] The incorporation of anti-war messages in the exhibitions of the Imperial War Museum reached its zenith during the early 1930s. One reason for this was the museum's relocation in 1935 to the former-Bethlem Royal Hospital building, Southwark. An erstwhile mental asylum, the building augmented its 'war to end all war' rhetoric with the idea that war equalled insanity.[65] Martin Conway, Lord Conway of Allington, the museum's first Director-General, implied this view when he participated in a House of Lords debate on the museum's relocation, declaring: 'I do not think that a lunatic asylum is at all a bad place for a War Museum'.[66]

Another reason was the prospect of further European-wide war which fuelled public revulsion and anti-war sentiment over the 1920s and 1930s across the United Kingdom.[67] Ironically, this occurred against renewed interest in the museum from the public who sought information about what might be expected from another armed conflict. The government also conducted urgent war research using the museum's collection. Responding to this use, Leslie Ripley Bradley, the museum's second Director-General and the one who would steer it through the Second World War, lamented in the 1938–1939 annual report that:

> [I]t cannot be too strongly emphasised that such are not the functions which the Museum was founded to perform, but rather was it to show the futility of war, and that its heroism is bought at all too dear a cost. It was to make an historical record of the war 'that was to end war', and not of the first of a series of world wars, each more terrible than the last.[68]

In reality, this line was maintained despite the original concept having already been entirely disproven following the waging of various brutal and protracted conflicts elsewhere.[69]

An observance of the 'war to end all war' myth by the Imperial War Museum was not maintained through ignorance or blind obeyance. Society must have granted it new meaning. This is plausible because understanding of concepts often accumulates—or loses—nuance over time. As Stuart Hall explains: 'meaning does not inhere in things [...]. It is constructed'.[70] Through public interaction, the 'war to end all war' myth altered depending how people conceived the world.[71] Malvern's analysis of the evolving societal reception to King George V's remarks at the museum's opening ceremony, which hinged on the understanding that the First World War was the 'war to end all war', supports this premise:

> Just after 1918, it could reinforce a mood of national mourning but it could also make criticism of the war seem like a dishonouring of the dead. By the mid 1930s, the statement might have been read as a powerful desire to avoid another war.[72]

It can be posited therefore that by April 1939, the myth of the 'war to end all war' had held various nuanced meanings since being invented. Initially, during the war itself, the words may have pushed messages intended to galvanise the population into defeating the Central Powers. Then, immediately after the war, it may have presented messages intended to help alleviate the nation's shock at what had happened. Next, as the country healed and families grieved during the 1920s, it may have communicated messages intended to reassure the bereaved. And finally, with another European-wide war looking likely during the 1930s, it may have conveyed messages intended to warn against history repeating itself. The Imperial War Museum drew on this cultural-cognitive institutionalisation. In doing so, it ensured the museum was viewed as congruent, rather than at odds, with the prevailing cultural beliefs.

The Imperial War Museum as an Organisation

Components of Organisation

Organisations can be understood as systemic entities established to help achieve specific aims and objectives.[73] Museums fit comfortably in this general understanding, possessing physical and non-physical commonalities with many other organisations.[74] As systemic entities, organisations possess manifold components. When combined and coordinated together, these components synergise to produce outcomes. There are multiple ways of theorising organisations. One effective approach is offered by John Child, who breaks organisations down into their physical and non-physical elements under three main categories—structural, processual and boundary crossing components—and analyses their roles.[75]

The first category comprises 'structural components'. These constitute the metaphorical flesh and bone of an organisation and are subdivisible into two further categories: structure and procedure. 'Basic structural components' allocate responsibility among organisational agents—personnel (staff)—along with the necessary physical and non-physical resources to see that responsibility met. 'Procedural components' set how an organisation acts and behaves, with such components including rules and standards, schedules and systems.[76] The second category comprises 'processual components'. Involving 'integration', 'control' and 'reward' measures, these enable an organisation to achieve the best productivity possible by aligning the cognitive and physical exertions of its workforce together.[77] And the third category comprises 'boundary crossing components'. Many organisations today regularly cross boundaries by letting their departments share remits, aims, objectives, and resources with other internal departments. Some even cross boundaries by outsourcing certain aspects of their operation to other organisations. This move supposedly lets the outsourcing organisation focus on their core business while conserving resources. Such behaviour was once rare in organisations. For example, the sharing of departmental aims, objectives and resources used to be considered uncontrollable, wasteful and chaotic, while outsourcing threatened organisational autonomy.[78]

Management and Structure

By the Second World War era, the Imperial War Museum had developed a five-tier hierarchical chain of management—not including department assistants—which became the basis for its basic structure. Figure 3.1 illustrates this structure which the museum took from 1933 until 1939. Such structures are often found underpinning museums, facilitating the so-called Classical Approach to organisational management developed by Henri Fayol.[79]

From 1936, the most senior staff member occupying the second tier was the Director-General, Leslie Ripley Bradley.[80] Responsibility for daily decisions made at the Imperial War Museum lay with him.[81] An alumnus of St John's College, Oxford, and an officer veteran of the First World War, Bradley joined the Imperial War Museum at its very beginning in 1917. He started working there as a storekeeper, managing the poster collection after being medically discharged from the British Army in 1916.[82] Over his career at the Imperial War Museum, Bradley climbed the ladder, gaining skills, knowledge, and experience; responsibility; and positions of authority. In 1933, he succeeded Charles ffoulkes as Curator and Secretary, and in 1936 Lord Conway as Director-General and Accounting Officer.[83]

A peculiarity of Bradley's employment at the Imperial War Museum is that when he became Director-General and Accounting Officer, nobody else assumed the roles of Curator or Secretary. Rather, throughout his tenure as Director-General, Bradley held all four positions simultaneously.[84] Together, these roles afforded him considerable discretionary decision-making capabilities and broad, direct responsibility. Indeed, drawing on Child's terminology, it eventually made him, as an agent, a 'critical structural component', with responsibility for setting the rules, systems and standards at the museum and for controlling their application. This explains the biographical quality the study sometimes attains and the occasions in the study where there is a general paucity of other voices represented.

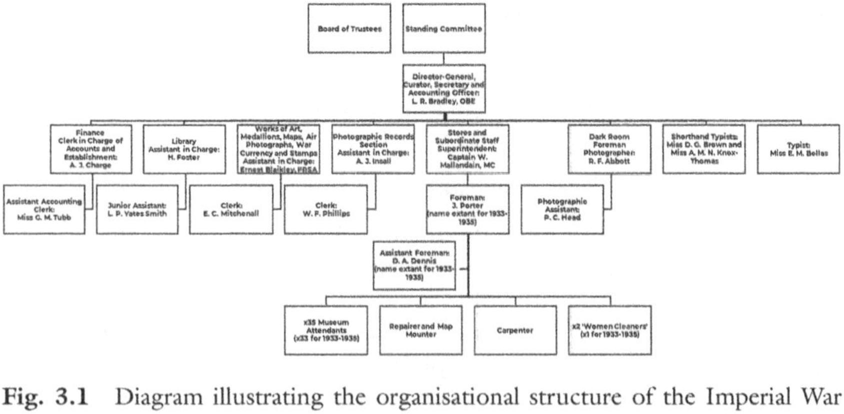

Fig. 3.1 Diagram illustrating the organisational structure of the Imperial War Museum, including staff names where these remain extant, 1933–1939

Staffing

Occupying the first tier of management at the Imperial War Museum, above Bradley, were the Trustees: 'a body corporate [...] with perpetual succession and a common seal, and power to acquire and hold land without licence in mortmain'.[85] As the museum's guardians, responsibility for setting its course, policies and agenda lay with them, their interests usually being represented by a Standing Committee.[86] This committee met four times during 1939.[87] Occupying the third tier, below Bradley, existed six departmental managers, each deputised by an assistant excepting the Superintendent.[88] These comprised specialist personnel focussed on the work of the department in which they were based.[89] And occupying the fourth and fifth tiers, below the Superintendent, were up to 56 staff comprising around 50 men and 6 women: on the fourth tier a Foreman and Assistant Foreman, and on the fifth tier Museum Attendants, a Repairer and Map Mounter, a Carpenter and two 'Women Cleaners' (an official job title).[90] These comprised support staff.[91] Their work involved performing schedules towards upkeeping and running the museum: invigilation, cleaning, maintenance, *etcetera*.[92]

The above structure removed an entire tier of middle management from the museum. On one hand, this would have provided departmental managers with greater agency (within parameters), responsibility and accountability over their work and decision making. But on the other, it effectively centralised and consolidated power around Bradley. It also ensured that he, as *the* Curator, had responsibility for integrating and coordinating the activities of the museum—this being a time when curators were generalists, rather than specialists.[93]

Flatter organisational structures can be very beneficial from an organisational perspective. It devolves decision making to the staff who carry out the work, fostering initiative and increasing accountability. It also breaks down stratified departmental boundaries, potentialising better internal collaboration between staff. But the process of transitioning from a lofty hierarchical structure to a flatter structure—usually by removing middle management—can prove costly from a non-monetary perspective, especially for museums. It could result in museums losing the long-developed knowledge, skills and experience possessed by the personnel, such as senior curators, who once held those positions. Hugh H. Genoways and Lynne M. Ireland warn that: 'These are staff members whose knowledge museums can ill afford to lose'. At the Imperial War Museum during the 1930s however, such concerns were less pronounced, owing to Bradley's retention by the institution and continued engagement in his various previous middle management roles.[94]

As with all comparable organisations, the Imperial War Museum incentivised its staff by rewarding them for their services that were rendered to realise the museum's aims and objectives.[95] The nature of the reward however depended on the individual staff member and their position in the museum. For the Trustees, this was the socio-cultural capital which holding such positions afforded them, a trusteeship being an entirely voluntary role.[96] For the staff, the basic reward was remuneration at varying rates. Senior staff could also be rewarded with civil service nomination, providing benefits such as greater job and pension security.[97]

Departments

The six specialist departments which Bradley oversaw at the Imperial War Museum were Finance; Library; Works of Art, Medallions, Maps, Air Photographs, War Currency and Stamps (this being one department); Photographic Records Section; Stores and Subordinate Staff; and Dark Room.[98] Excepting Stores and Subordinate Staff which employed the majority of the museum's staff, each department had just two personnel: a departmental manager—often referred to as an 'officer'[99] in the archived museum documentation—and an 'assistant'.[100] The museum's roster also mentions one typist and two shorthand typists without clear affiliation.[101] It is reasonable to assume that these typists made up a typing pool which worked across all of the departments, including with the Director-General, for very little bureaucratic documentation preserved in the museum archive has been handwritten.

Relationships

Over its development the Imperial War Museum had created a culture of fostering and maintaining relationships and interactions with other organisations. Such exchanges can be extremely beneficial to the museum ecology. They foster opportunities for bilateral exchange of knowledge, skills, experience and support.[102] Examples from the Imperial War Museum during the 1930s include those with the Standing Commission on Museums and Galleries, which advised the government on national museum and gallery matters; the Museums Association, which promoted professional standards in museums; and museums themselves: national museums such as the National Maritime Museum, provincial museums such as the London Museum, and those overseas such as the Australian War Memorial.[103] Interestingly, the directors of two other London-based national museums, the British Museum and the Science Museum, were also Trustees of the Imperial War Museum.[104] This arrangement may have been to compensate for the limited museological expertise there discussed below.[105]

Of equal importance were the relationships and interactions maintained by the Imperial War Museum with the government ministries and departments responsible for supporting the museum and/or appointing its Trustees. These ensured the museum remained operational. Two significant examples here were the relationships maintained with HM Treasury and the Office of Works, the latter which would become the Ministry of Works in 1940. Although the museum had officially been constituted as an independent 'Government Museum',[106] it was at times required to defer its autonomy to the Treasury.[107] In this sense, the Treasury effectively monopolised the museum, being responsible not only for providing the museum's funding, but also for appointing 11 of the museum's 24 Trustees.[108] The Office and later Ministry of Works, by contrast, maintained and developed the museum's premises. Its ministerial head also sat on the Board of Trustees in an *ex officio* capacity.[109] Other governmental ministries or departments which the museum maintained a close relationship with were the War Office, the Air Ministry, the Colonial Office and the India Office, whose ministerial heads appointed one Trustee each; and the Admiralty and the Board of Education, whose executive committees also had responsibility for appointing one Trustee each.[110]

One particular relationship fostered over the 1920s and 1930s provides an interesting example of organisational boundary crossing discussed above. As previously remarked, this was unusual at the time.[111] On being formally established in 1920, the Imperial War Museum became responsible for the care and preservation of the country's official war cinematograph films—a new and exciting yet hazardous technology which could be extremely flammable. Not possessing any staff with the skills to undertake such work, the museum was required to outsource responsibility to the government's cinematography advisor, Edward Staple Foxen Cooper, then based at the Foreign Office. This arrangement quickly evolved. After several months, the museum formally employed Foxen Cooper on a part-time basis, shared with the Foreign Office and HM Customs and Excise, until he retired in 1934.[112] No documentation has been found to indicate whether any replacement was sought thereafter. It suggests Foxen Cooper's skills had been absorbed by the museum over his tenure. This is posited because after Bradley retired as Director-General in 1960, the Trustees retained his services on a part-time basis to rationalise the museum's film collection, work which required those skills.[113]

Public Interaction

Much the same can be written about the Imperial War Museum and the public at large. Integral to the museum's public profile and legitimacy were the relationships it fostered with constituency groups. Indeed, there are numerous examples recorded in documentation held by the museum archive of onsite interaction between the museum and service associations, schools and youth organisations—people which often visited the museum for instructional and/or recreational purposes. Eager to be used, the museum became receptive to requests for assistance by core and other constituency groups which might have seen that outcome realised, and would try to support them where it could.[114] Museums can often seem like unattractive, hostile or irrelevant amongst different communities.[115] By engaging with various different constituency groups such as schools and service associations, the Imperial War Museum could potentially extend its reach to previously unengaged members of the public.

Alongside establishing and maintaining relationships with constituency groups, the Imperial War Museum also developed an ethic of interacting with individual users. It did this by responding routinely to postal and telephone enquiries.[116] The voluminous collection of preserved public correspondence for the period is fascinating and comprises a remarkable source of social history. They convey information about people and their wide-ranging emotions that often cuts through the bureaucratic documentation otherwise making up the museum archive. What is so striking from reading them and the responses is the extent to which the museum would reply with personalised messages. Even the most mundane and monotonous enquiry, such as those requesting the museum's opening times and facilities, were usually answered by the Director-General himself (or at least in his name).

The Imperial War Museum as a Museum

Museum Collections and Exhibitions

Museums are diverse entities which variously collect, preserve and present the culture of society and its environment. This means that no museum holds identical characteristics with another. But when stripping back any museum to their essential features, two commonalities can usually be observed. The first is a collection of heritage assets acquired from the social system.[117] The second is a series of exhibitions which present those assets to the public.[118]

As the appointed or indeed self-appointed caretakers of society's history and heritage, museums collect and preserve material that represent the lived environment, both social and scientific. These objects form the basis for museum collections, which museums hold in trust. There are multiple categories of objects that comprise museum collections. Suzanne Keene identifies four main examples applied in this study.[119] These include 'objects for visual enjoyment', 'functional objects', 'objects for research', and 'place-' and 'people-based' objects. A recently arising fifth category concerns 'virtual objects', however these are not considered herein owing to the study's historical setting.[120] Few collections comprise just one typology. Many museums, no matter their remit, possess material drawn from all four.

Objects for visual enjoyment are those collected to please the eye. These often feature in fine and decorative art collections, but could feature in any collection.[121] Functional objects, by contrast, are those collected to demonstrate practical use or design. These usually feature in industrial collections of machinery such as boats, vehicles or aeroplanes, or design collections of fine pieces of precision engineering such as time pieces and instruments.[122] Objects for research, by contrast again, are primarily those collected to be studied rather than exhibited. Such objects often comprise archives or *de facto* archives, which raises important questions about what comprises an object and whether an accession of multiple pieces comprises one object or more than one.[123] In this study, an object is conceived as any sole or assembled item that has been selected to represent something specific. And finally, place- and people-based objects, by contrast again, are those collected because they support a museum's *raison d'être* and rationale. These often are the bulk of most museum collections.[124] It is important to emphasise that an object will not necessarily embody just one typology. A place- and people-based object could also comprise a functional object, a functional object could also comprise an object for visual enjoyment, and an object for visual enjoyment could also comprise an object for research. This potential overlap means some objects could embody three typologies or more, with any distinction greatly depending on subjective perception.

Museum exhibitions help museums fulfil their role as caretakers of society's history and heritage. They are key markers in determining a museum's performance and regularly subject to review by critics, professionals and journalists.[125] Through producing exhibitions, museums can showcase responsible stewardship of collections. They can also facilitate public access to them at varying levels and demonstrate their productivity. Each of these are important for demonstrating institutional and organisational legitimacy.[126]

Museum exhibitions can be effective and efficient communicative mediums.[127] They can also be outlets for great ingenuity and creativity.[128] Obviously, success or failure in achieving either attribute depends greatly on their design quality. A design not only sets out an exhibition's presentation, but is also key in crafting an exhibition's message, and the method by which an exhibition's message gets conveyed. There are thus many conceivable considerations that must be engaged with during the exhibition design process. These include the exhibiting museum's mission, values and vision; the available space; the environment; the narrative being told; the objects identified for display, their contribution to the narrative and arrangement in the space; the methods of interpretation; lighting, sound and other sensory attributes; amongst many conceivable others.[129] All will greatly impact on an exhibition's dynamic, with any imbalances potentially impacting the visitor experience.

This observation about the fundamentality of collections and exhibitions to museums is far from a novel idea. Between 1889 and 1939, the museum sector in the United Kingdom had mutated into one with basic professional attributes and standards clearly recognisable across institutions today.[130] Progress was expedited by the publication of three watershed reports on museums and museum practice—the Miers Report of 1928; the Royal Commission on National Museums and Galleries reports of 1928, 1929 and 1930; and the Markham Report of 1938[131]—which recommended improvements to national and provincial museum facilities, exhibition quality and variety, and opening hours (especially for provincial museums).[132]

Each report identified collections and exhibitions as being key elements of museums. Particularly, the Miers and Markham reports were at pains to point out the importance in fostering rational collections and good quality exhibition practices. The former did this by stating: 'Museums may be regarded as existing for the purpose of storing, exhibiting and utilising objects of cultural and educational value'.[133] And the latter did so by designating the collection and preservation of objects and, what it called, the development of 'visual education' as being the first and third great functions of museums respectively—the second being research.[134] The report also noted that the third function was 'now growing so greatly important that it completely overshadows the other two in the public mind'.[135]

Collection

As an active museum on the First World War, the Imperial War Museum admitted new place- and people-based objects over the inter-war years specifically from or concerning the events of 1914–1918. Accessions ranged from *matériel* and items from social life to artwork and books. It also collected functional objects, objects for visual enjoyment, and objects for research. Information on each accession can be found in the annual reports, with articles about the most interesting being published in the *Museums Journal*.[136] But the extent of this inter-war collecting never equalled that undertaken by the organisers during the museum's founding period spanning 1917–1920. This is because most material held at the Imperial War Museum throughout the inter-war years was largely gathered via a formal contemporary collecting programme before it had opened. Such programmes are instigated by museums to collect items in the moment, usually during some protracted event or phenomenon.[137] They secure the raw materials which later inform history for future generations.[138] Such formal contemporary collecting programmes benefit museums because, as Owain Rhys and Zelda Baveystock explain, amongst other reasons, they 'capture a fuller more nuanced record of society whilst material is abundantly available'.[139] This includes capturing the provenance and other contextual information surrounding the collected objects—the stories associated with them, which transform specimens into objects with an interesting and informative past (historically, a concern of the Imperial War Museum)[140]—that becomes murkier the further the collector is from the event.[141] Kavanagh argues that the organisers' efforts during its early years resulted in the most comprehensive collection work undertaken towards any British museum and that their achievements have at times been underappreciated.[142]

The organisers of the Imperial War Museum had ambitious aspirations for its collection. For example, Sir Martin Conway, the first Director-General, had expressed early on in the organising phase his desire that they collect 'not only every type of gun, but a type of each gun in the various stages of its development'.[143] To assume from this stated hope however that the organisers envisaged a museum concerned purely with preserving and displaying *matériel* would be incorrect. By 1917, the United Kingdom was fighting the First World War on the front line and home front. In doing so, the conflict had transitioned into a 'total war': where an entire country's population and resources is engaged somehow with the war effort towards the enemy's total capitulation.[144] The organisers therefore aspired not just to make the museum's collection representative of the British military's and later its colonial militaries' exploits, but also the United Kingdom and British Empire's total experience.[145]

This far-reaching ambition is demonstrated by the considerable number and variety of collecting subcommittees that the organisers established in early 1917. They initially comprised the Admiralty, munitions, records and literature, Red Cross, War Office, and women's work subcommittees. Later that year the Red Cross subcommittee was dissolved and the air services and dominions subcommittees created. The War Office subcommittee never really formed and eventually became superseded by the War Office's own War Trophies Committee, which transferred over to the museum's organisers. During 1918, an Art Committee was also formed. Additionally, a Canadian Army officer called Major Henry Beckles Willson, who had prior experience of staging war exhibitions for the Belgian and Canadian relief funds, received employment gathering suitable artefacts from Belgium and France.[146] Although collection items were slow to amass at first, the various organisers prosecuted their tasks with enthusiasm. Their persistence paid dividends. Through establishing official arrangements with government departments and ministries, and by making repeated public appeals, material started accumulating in the makeshift stores.[147]

Such was the intended completeness of the collection at the Imperial War Museum that the organisers sought material which previously would have been considered worthless, but which now is considered fundamental in representing social history.[148] This has led Malvern to consider its collection today as 'an ethnographic collection for the display of the British nation-in-arms'.[149] Reflecting on writing by Susan M. Pearce, Paul Cornish however notes that the systematic collecting required in achieving the organisers' full aspirations was unrealistic.[150] This is supported by the fact that many ephemeral items from 1914–1915 remained uncollected by the Second World War years, as discussed later during Chap. 6.

The organisers of the Imperial War Museum were not only interested in functional objects and objects for study. As already discussed, they also sought objects for visual enjoyment. In fact, some of the more renowned items acquired over its foundation were artworks. This decision meant that the Imperial War Museum developed into an extensive modern art gallery as well as a museum on war and armed conflict.

The art collection originated from an initiative commenced in 1916 by the state's propaganda bureau, Wellington House.[151] This came about owing to public demand for visual representations of the war coupled with the impending onset of conscription, the military's hostility towards photography near the frontline, and perception of a dearth of 'interesting art' in the capital's art institutions.[152] The following year, an agreement was struck between the organising committee and the Department of Information—which replaced Wellington House—establishing that the propaganda collection would be transferred to the Imperial War Museum after the war.[153] But this came into doubt in 1918 when the Ministry of Information, headed by a known opponent of the museum project called William Maxwell Aitken, 1st Baron Beaverbrook, replaced the Department of Information and established its own British War Memorials Committee for commissioning commemorative war art.[154] Thus, to avoid ending up with no art at all, the organising committee initiated its own collection.

The organising committee commissioned eyewitness paintings and drawings through its various subcommittees.[155] Whereas the propaganda/memorial collection focussed on art that represented the Western Front, the Imperial War Museum collection, in contrast, focussed on art representing the naval and air war. It also included art representing women's war work, plus detailed technical drawings.[156] In the end however, fears over Beaverbrook's hostility proved unwarranted, as the museum did end up receiving the propaganda/memorial collection.[157] This left the Imperial War Museum, by 1919, with no fewer than 3000 individual artworks.[158] Malvern argues that it resulted in the Imperial War Museum during the inter-war years holding 'the most significant and important collection of modern British art in the country': 'more comprehensive than the holdings of the Tate Gallery'.[159]

Exhibitions

From the very beginning, the Imperial War Museum used its collection to create informative and emotive exhibitions. Reviews suggest that the sight of the material on display succeeded in producing dynamic, sensory experiences.[160] At the Crystal Palace, a free flowing, open plan gallery accommodated three central, principal exhibitions concerning the army, navy and air force surrounded by various separate, smaller displays enclosed within their own alcoves. Those smaller displays, concerning themes pertaining to the army, navy, or air force, such as the merchant navy, anti-aircraft defences and trench warfare, were positioned alongside their respective central exhibition. Those which concerned themes without clear linkages, or themes considered less important, were displayed on the peripheries, examples here being the medical services, veterinary section and women's work.[161] The historical objects located in these exhibitions often shared their space with artworks. Art had an important interpretive function at the Imperial War Museum. While some pieces—usually from the propaganda/memorial collection—were hung separately in dedicated 'art galleries', others—such as those commissioned by the museum—were hung alongside the same or similar *matériel* to help illustrate their functionality.[162] As a result, Malvern suggests that those latter examples inhabited more of an interpretative rather than an aesthetic function: a practice which still takes place at the museum today when photographic illustrations elude curators.[163] This general description represents the form that the museum's exhibitions took over the first forty years.[164] It is also arguably representative of the level of museum training and technical abilities of the staff employed there, these being mostly ex-service personnel, not trained curators. Until Noble Frankland's tenure as Director-General in 1960, the museum consciously chose to employ ex-service personnel for their direct war experience.[165]

The exhibitions at the Imperial War Museum drew significantly on the lives and experiences associated with the historical objects they presented. This is because many of the exhibits, such as ephemera and equipment, often comprised one of thousands of identical examples in a series. Indeed, what made the museum's particular specimens so significant compared with many other examples was their provenance. Accordingly, a desire to secure reliable provenance steered the organisers' collecting.[166] Thus the objects, and even the exhibitions and museum itself, became conduits for stories. This idea has previously been alluded to in an evocative animated marketing video released by IWM titled *Flight of the Stories*, where double quotation marks (“) are depicted flying like birds away from the battlefields of Europe, over the English Channel, towards London and the museum, eventually resting outside on its iconic dome, roof, naval guns and gardens.[167] Exhibits holding verified stories, especially with gravitas or perceived importance, were sometimes exhibited separate from the main displays. This approach, as identified by Cundy, highlighted the artefacts' historical significance and boosted their affective resonance.[168] In the case of the gun from L Battery discussed above, she explains its potential to induce commemorative responses was harnessed through emphasising the 'loss and heroism' surrounding the object, this quality having been drawn out by the object's spatial proximity to the cenotaph.[169]

When the Imperial War Museum moved to the Imperial Institute in 1924, the limited space available there prevented the museum from expanding. But it did not stifle a creative curatorial mindset. In fact, the spatial limitations prompted the museum to think even more creatively about how ideas and objects could be interpreted through the exhibition format. One such example at this time can be seen through its innovative use of moving images. The museum employed two Mutoscopes to display various flick book style films, with nine programmes available. Roger Smither and David Walsh believe this to be the first occasion where moving images have been used as an interpretive device in any museum setting.[170] Another example can be seen with the reconstruction of a trench dugout using material scavenged from local waste ground.[171] Today, both interpretive devices are routinely used in museum exhibitions worldwide for their ability to inform and engage audiences.[172]

As discussed above, the exhibitions of the Imperial War Museum had changed little over its three pre-Second World War presentations. At the former-Bethlem Royal Hospital building, material continued to be arranged around the three core themes of the army, naval and air services.[173] Figure 3.2 visualises the building's layout at this time, while Figs. 3.3, 3.4 and 3.5 depict these galleries soon after its transfer there. The collection and display policies also remained largely unaltered, meaning that the exhibitions presented themes and material from a strictly limited period. Yet the exhibitions were not stagnant. The three annual reports preceding the Second World War portray a year-on-year process of modification across the displays.[174] For example, during the financial year 1938–1939, the museum made no fewer than 16 alterations. Most seem to be little more than attempts at enriching the exhibitions through adding two- and three-dimensional objects of various object types. Some, however, appear to have ulterior motives, representing the darkening national picture: an aggressive neighbouring power and a rapid national rearming in response.

One such example is the change made to the German Army Gallery. This comprised the addition of a rectangular ochre wooden shop shutter taken from Metz-en-Couture in France by British soldiers on 4 April 1917. Across the shutter was a derisory and racist poem, painted in English with black lettering, by somebody from the withdrawing German forces:

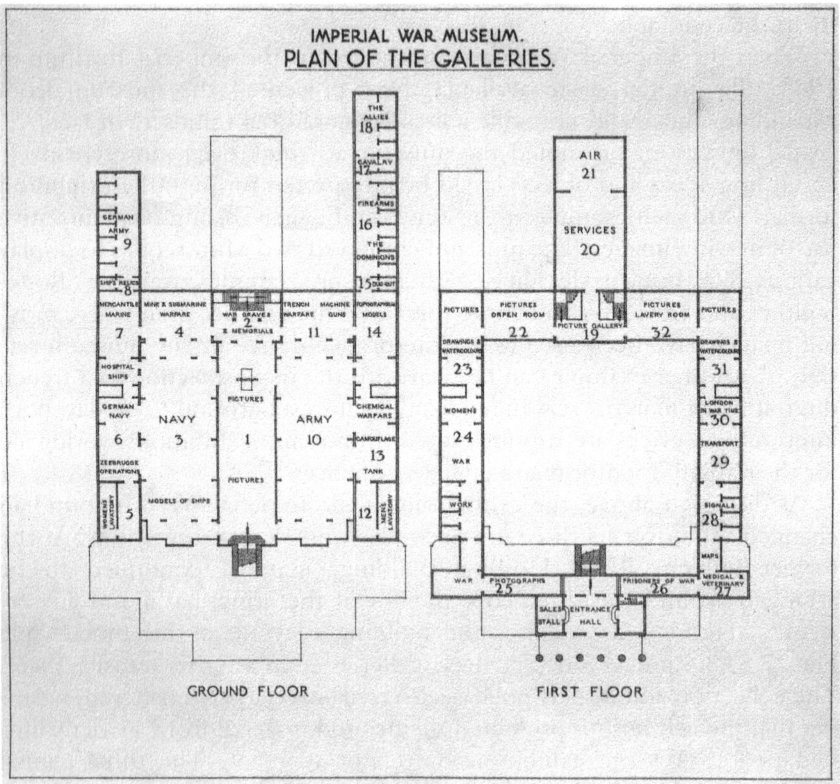

Fig. 3.2 Plan visualising the gallery spaces of the Imperial War Museum, 1936–1939. © IWM (Q 60569)

Fig. 3.3 Photograph depicting the Army Gallery, c. 1937. © IWM (Q 61184)

> You crie [*sic*]: Poor little Belgium!
> Poor Ireland you don't care,
> Protecting culture God and law
> You brought the n[…] there.
> I know you're always hypocrites
> Now hear, what I you tell:
> Our Germany will go to head
> But you oh, go to hell!!
> With every good wish for a Happy Xmas
> Und bright New Year at Metz e.C.
>
> Yours truly, German.[175]

Another is change which occurred in the Prisoners of War and War Souvenirs Gallery, where recruitment adverts from the First World War were added and displayed on the walls. The material included posters such as 'Remember Belgium: Enlist Today' and 'Daddy, what did you do in the

Fig. 3.4 Photograph depicting the Naval Gallery, c. 1937. © IWM (Q 61183)

Great War?' by printer Henry Jenkinson and artist Savile Lumley respectively. Also added was the original black and white paint and graphite picture by Alfred Leete that adorned the front cover of 5 September 1914 edition of the *London Opinion*, famously depicting Britain's Secretary of State for War from 1914 to 1916, Horatio Herbert Kitchener, 1st Earl Kitchener, with 'Your Country Needs You' blazoned beneath him.[176]

Of course, the rationale for these amendments can only be speculated on without first-hand curatorial insight.[177] But rationale is not what makes them significant here. Rather, their significance lies in what the amendments apparently indicate about the attitude of the Imperial War Museum towards its exhibitions at the time. Firstly, they suggest that the museum, despite maintaining a static collection and exhibition policy, was concerned with ensuring its exhibitions were enhanced, freshened up, and curated so the issues and ideas represented bore relevance to current affairs. Secondly, they imply that the museum did not shy away from

Fig. 3.5 Photograph depicting the Air Services Room, c. 1937. © IWM (Q 61377)

addressing provocative issues. At a time when many of its visitors would have been anti-war, displaying recruitment posters and an exhibit inciting anti-German sentiments was bold if not provocative. And thirdly, they show that the museum approached curation pragmatically. Despite protests by the Director-General in 1939 complaining about public use of the museum for purposes opposed to its anti-war agenda, the museum nonetheless responded by satisfying this curiosity through updating the existing exhibitions accordingly.

Conclusion

Over its first 19 years of operation, the Imperial War Museum developed into a purposeful institution, organisation and museum. Resulting from an initiative that was part propaganda, part sincere attempt to document the world war fought over 1914–1918, it became, effectively, an important centre and mnemonic space for the study and remembrance of the 'war to end all war'. The museum achieved this status by performing two tasks. The first was to inform the public about the conflict. The second was to help commemorate those who had fought and died in the conflict from the homeland, dominions and empire. Through the latter, the museum became a *de facto* centre for national and imperial commemoration.

Initially a regulative institution to reinvigorate the British national war effort, the Imperial War Museum soon became a normative institution, profiling the ensuing sacrifice and instilling reverence in remembrance of national and colonial sacrifice. Simultaneously, it increasingly embodied a cultural-cognitive institution as the museum increasingly represented society's developing views about the 'war to end all war'. Accordingly, from comprising an instrument for furthering war before morphing into an institution that conciliated victim families in its aftermath, the museum transitioned into an institution which admonished war, becoming, what Joyce Apsel has come to conceive, an early peace museum, where the benefits of peace were shown by graphically depicting the horrors of war.[178]

Over the inter-war years the Imperial War Museum developed into an organisation with an unusual personnel management structure. Instructions from the Director-General descended directly on departmental management facilitated by a limited hierarchy. This meant more staff members than usual could be responsible for interpreting museum policy. The staff themselves were mainly veterans of the armed forces. While often inexperienced in museum work, demonstrated through their simplistic approach to exhibition design, they nevertheless told affective stories by conceiving and implementing innovative museological ideas and muesographical practices using space, objects and text in distinctive ways.[179]

The Imperial War Museum also developed at this time into a museum with distinct organisational architecture. Excepting virtual organisation, which did not become possible until the 1990s, every component outlined through Child's conceptualisation of organisation is observable from bureaucratic documentation preserved in its archive.[180] A public body, the museum, as with most established organisations, had basic structure, operated on procedures and processes, and maintained relationships with other organisations, especially government departments and ministries. The most important of these were the relationships with government departments and ministries which presided over, maintained, and facilitated the museum's work. Unusually for a historical organisation, it even crossed organisational boundaries. In many ways therefore, the essential features of the museum's organisational architecture are similar to those held by present organisations generally, even if the technology incorporated now has developed since the early mid-twentieth century.

Collecting and exhibitions were as essential to the Imperial War Museum in the inter-war years as today. Its collection comprised heterogeneous material from the front line and the home front, amassed predominantly over 1917–1920. The diversity of the material afforded the collection an ethnographic quality, comprising objects from each object typology identified by Keene. In establishing the collection, the museum became a modern art gallery as well as a war museum, because the material not only composed historical objects but also artwork. Moreover, its exhibitions were simple yet effective. The museum managed to produce exhibitions that conveyed messages through creative approaches to interpretation. This included employing new technologies and experimenting with object placement.

Following 1933, the increasingly unavoidable threat to the Imperial War Museum was its increasingly anachronistic subject remit: the world war of 1914–1918 as the 'war to end all war'. This arose through being founded on static collection and exhibition policies which bounded collecting and exhibitions within the years 1914–1918 only. Fortuitously, the museum perceived its increasing anachronisation as the potential for another war increased. The existing exhibitions were therefore given subtle changes which reflected this situation. Indeed, while paying lip service to its founding mission, the museum started departing from the philosophy underlying it by subtly acknowledging the growing threat of another war.

Through drawing the above conclusions, the current chapter makes the two following essential points in support of the overarching study. One is that during the inter-war years, the Imperial War Museum established a precedent for institutional, organisational and museological adaption. It acquired this precedent through responding reflexively to societal change as society underwent value shifts, a characteristic which the museum carried into the Second World War. Hence, the chapter also shows that by 1939 the museum was not unfamiliar with adjusting to short-term pressures or long-term societal developments, a quality which would come in useful when managing the challenges posed by the Second World War. The other point is that the case bears many similarities with present-day museums. This legitimises the historical approach taken to the study. Despite focussing on a situation which is, at time of writing, more than

80 years old, there are multiple institutional, organisational and museological issues and ideas raised which still resonate today—and will do so still for the foreseeable future. Accordingly, the current chapter underlines this study's applicability to museums facing crisis and change in the prevailing context.

The next chapter continues the theme of adaption. It examines ways in which the Imperial War Museum resolved disruption to what is conceptualised as the institution's civic service provision during the Second World War. This crisis-conducive situation impacted all British cultural and creative institutions, but particularly the London national museums and galleries. Through doing so, the chapter expands on the concept of organisational resilience introduced during Chap. 2 and explores its role in managing and resolving crises.

Notes

1. Otto Lerbinger, *The Crisis Manager: Facing Disasters, Conflicts, and Failures*, 2nd edn (New York, NY: Routledge, 2012), 22; Montgomery Van Wart, 'The First Step in the Reinvention Process: Assessment', *Public Administration Review* 55 (1995): 429–430.
2. Nassim Nicholas Taleb, *The Black Swan: The Impact of the Highly Improbable* (London: Allen Lane, 2007).
3. Mara Cerquetti and Eleonora Cutrini, 'Structure, People and Relationships: A Multidimensional Method to Assess Museum Resilience', *Nonprofit and Voluntary Sector Quarterly* 52 (2023): 135.
4. Douglas North, *Institutions, Institutional Change and Economic Performance* (Cambridge: Cambridge University Press, 1990), 3–4; W. Richard Scott, John W. Meyer and associates, *Institutional Environments and Organizations: Structural Complexity and Individualism* (Thousand Oakes, CA: SAGE, 1994), 68.
5. W. Richard Scott, *Institutions and Organizations: Ideas, Interests and Identities*, 4th edn (Thousand Oakes, CA: SAGE, 2014), 57.
6. Andrew J. Hoffman, *From Heresy to Dogma: An Institutional History of Corporate Environmentalism*, exp. edn (Stanford, CA: Stanford Business, 2001), 36.
7. Scott, *Institutions and Organizations*, 59, 70.
8. Ibid., 62.
9. Diana M. Condell, 'The Imperial War Museum, 1917–1920: A Study of the Institution and Its Presentation of the First World War' (master's thesis, n.place, 1985), 15.

10. Gaynor Kavanagh, *Museums and the First World War: A Social History* (London: Leicester University Press, 1994), 121–122.
11. George H. Cassar, *Lloyd George at War, 1916–1918* (London: Anthem, 2009), 19–74.
12. David Monger, *Patriotism and Propaganda in the First World War: The National War Aims Committee and Civilian Morale* (Liverpool: Liverpool University Press, 2012), 24–26.
13. Kavanagh, *Museums*, 122.
14. Monger, *Patriotism and Propaganda*, 17.
15. Diana Condell, 'The History and Role of the Imperial War Museum', in *War and the Cultural Construction of Identities in Britain*, ed Barbara Korte and Ralf Schneider (Amsterdam: Rodopi, 2002), 29.
16. Robin Prior and Trevor Wilson, *The Somme* (New Haven, CT: Yale University Press, 2005), 300–301.
17. Peter Simkins, 'The Four Armies, 1914–1918', in *The Oxford History of the British Army*, ed David G. Chandler with the assistance of Ian Beckett (Oxford: Oxford University Press, 1996), 240–242.
18. Condell, 'History and Role', 29.
19. Kavanagh, *Museums*, 117–119.
20. Condell, 'Imperial War Museum', 14–15.
21. Ibid., 59–64.
22. Scott, *Institutions and Organizations*, 59–64; Hoffman, *Heresy to Dogma*, 36.
23. Scott, *Institutions and Organizations*, 66–70.
24. Kavanagh, *Museums*, 36–44.
25. Anthea Lang, 'Shipley Bequest: Joseph Shipley and His Art Gallery' ([Gateshead?]: privately produced, n.d.).
26. Kavanagh, *Museums*, 122.
27. Jonathan Conlin, *The Nation's Mantlepiece: A History of the National Gallery* (London: Pallas Athene, 2006), 50.
28. *Imperial War Museum Act*, 1920, c. 16; IWM, 'Annual Report and Accounts, 2022–2023' ([London?]: HMSO, 2023), 60, https://assets.publishing.service.gov.uk/government/uploads/system/uploads/attachment_data/file/1178402/Imperial_War_Museum_Annual_Report_and_Accounts_2022_to_2023.pdf (18 March 2024); Alfred Mond, *Parliamentary Debates*, Commons, 5th ser., vol. 127 (1920), col. 1465–1467.
29. *Parliamentary Debates*, Commons, 5th ser., vol. 103 (1918), col. 1407.
30. Kavanagh, *Museums*, 136.
31. *National Maritime Museum Act*, 1934, c. 43; William Ormsby-Gore, *Parliamentary Debates*, Commons, 5th ser., vol. 291 (1934), col. 1495–1499.

32. Scott, *Institutions and Organizations*, 60, 63.
33. Kavanagh, *Museums*, 140–143.
34. Alyson Mercer, 'The Changing Face of Exhibiting Women's Wartime Work at the Imperial War Museum', *Women's History Review* 22 (2013): 335–336.
35. Kavanagh, *Museums*, 131, 135–136.
36. Ibid., 136.
37. Scott, *Institutions and Organizations*, 64–66.
38. Hoffman, *Heresy to Dogma*, 36.
39. Scott, *Institutions and Organizations*, 64–66.
40. Ibid., 60.
41. Ibid.
42. 'The King Opens War Museum', *Daily Mail*, 10 June 1920; 'King and Queen at the War Museum', *Daily Telegraph*, 10 June 1920; 'Imperial War Museum', *Manchester Guardian*, 10 June 1920; 'The Greatest War Memorial', *Times*, 10 June 1920.
43. Sue Malvern, 'War, Memory and Museums: Art and Artefact in the Imperial War Museum', *History Workshop Journal* 49 (2000): 185.
44. IWM, Museum Archive (MA), EN1/1/REP/003, 'Third Annual Report of the Imperial War Museum, 1919–1920', 3.
45. Ibid., 4.
46. Malvern, 'War', 185–188; Steven Cooke and Lloyd Jenkins, 'Discourses of Regeneration in Early Twentieth-Century Britain: From Bedlam to the Imperial War Museum', *Area* 33 (2001): 385; Stefan Berger, Anna Cento Bull, Cristian Cercel, David Clarke, Nina Parish, Eleanor Rowley and Zofa Wóycicka, 'Memory Cultures of War in European War Museums', in *Agonistic Memory and the Legacy of 20th Century Wars in Europe*, ed Stefan Berger and Wulf Kansteiner (Cham: Palgrave Macmillen, 2021), 69–114; Kasia Tomasiewicz, '"We are a social history, not a military history museum"': Large Objects and the "Peopling" of Galleries in the Imperial War Museum, London', in *Museums, Modernity and Conflict: Museums and Collections in and of War Since the Nineteenth Century*, ed Kate Hill (Abingdon: Routledge, 2021), 219.
47. Jennifer Wellington, *Exhibiting War: The Great War, Museums, and Memory in Britain, Canada, and Australia* (Cambridge: Cambridge University Press, 2017), 244.
48. James Scott, 'Objects and the Representation of War in Military Museums', *Museum and Society* 13 (2015): 490–493.
49. Paul Cornish, '"Sacred Relics": Objects in the Imperial War Museum, 1917–1939', in *Matters of Conflict: Material Culture, Memory and the First World War, ed Nicholas J. Saunders* (Abingdon: Routledge, 2004), 46.

50. Jan Geisbusch, 'For Your Eyes Only? The Magic Touch of Relics', in *Museum Objects: Experiencing the Properties of Things*, ed Sandra H. Dudley (Abingdon: Routledge, 2012), 202.
51. Wellington, *Exhibiting War*, 244–247.
52. IWM, MA, EN1/1/REP/005, 'Fifth Annual Report of the Imperial War Museum (Second Report of the Trustees), 1921–1922', 2.
53. Alys Cundy, 'Objects of War: The Response of the Imperial War Museum, London, to the First and the Second World Wars', *Post-Medieval Archaeology* 51 (2017): 266.
54. Alys Cundy, 'Thresholds of Memory: Representing Function through Space and Object at the Imperial War Museum, London, 1918–2014', *Museum History Journal* 8 (2015): 260.
55. Michel Foucault, 'Of Other Spaces', trans. Jay Miskowiec, *Diacritics* 16 (1986): 22–27.
56. Cundy, 'Thresholds of Memory', 254.
57. Ibid., 254–255.
58. John Hulton, *The Gunners of 1914: Baptism of Fire* (Barnsley: Pen and Sword Military, 2014) 87–88.
59. Cundy, 'Thresholds of Memory', 257–258.
60. Saadi Lahlou, *Installation Theory: The Societal Construction and Regulation of Behaviour* (Cambridge: Cambridge University Press, 2018).
61. Malvern, 'War', 188–189.
62. Cooke and Jenkins, 'Discourses of Regeneration', 385.
63. David Olusoga, *Black and British: A Forgotten History*, pbk edn (London: Pan, 2017), 427–466.
64. Scott, *Institutions and Organizations*, 60, 66–70; Hoffman, *Heresy to Dogma*, 36.
65. Malvern, 'War', 192.
66. Martin Conway, *Parliamentary Debates*, Lords, 5th ser., vol. 86 (1932), col. 152.
67. Richard Davis, 'The British Peace Movement in the Interwar Years', *Revue Française de Civilisation Britannique* 22 (2017): 10.4000/rfcb.1415, https://journals.openedition.org/rfcb/1415 (18 March 2024).
68. IWM, MA, EN1/1/REP/032, '19th Report of the Director-General and of the Curator and Secretary to the Board of Trustees for the Years 1935–1936 and 1936–1937', 5.
69. Philip Ziegler, *Between the Wars* (London: MacLehose, 2016).
70. Stuart Hall, 'The Work of Representation', in *Representation*, ed Stuart Hall, Jessica Evans and Sean Nixon, 2nd edn (London: SAGE, 2013), 10.
71. Scott, *Institutions and Organizations*, 67.
72. Malvern, 'War', 181.

73. North, *Institutions*, 3–5.
74. William M. Sukel, 'Museums as Organisations', in *Museum Management*, ed Kevin Moore (London: Routledge, 1994), 262–264.
75. John Child, *Organization: Contemporary Principles and Practice* (Malden, MA: Blackwell, 2005), 6–17.
76. Ibid., 6–7.
77. Ibid., 8, 12–13.
78. Ibid., 8–9, 15–17.
79. Hugh H. Genoways and Lynne M. Ireland, *Museum Administration: An Introduction* (Walnut Creek, CA: AltaMira, 2003), 40–41; Michael A. Fopp, *Managing Museums and Galleries* (London: Routledge, 1997), 11–15.
80. IWM, MA, EN1/1/REP/032, see reports therein.
81. Steven Miller, *Anatomy of a Museum: An Insider's Text* (Hoboken, NJ: Wiley Blackwell, 2018), 29–44.
82. The National Archives (TNA), WO 339/54103, Davies to Bradley, 29 September 1916.
83. 'Mr L. R. Bradley', *Times*, 30 January 1968.
84. IWM, MA, EN1/1/REP/032, see reports therein.
85. *Imperial War Museum Act*, 1920.
86. Child, *Organization*, 6, 9, 11; IWM, MA, EN1/1/REP/032, see reports therein; IWM, MA, EN2/1/ACC/004/7, Forsdyke to Bradley, 5 February 1941.
87. IWM, MA, EN1/1/COB/049, see meeting minutes therein for the four meetings held over 1939.
88. IWM, MA, EN1/1/REP/032, see reports therein.
89. Child, *Organization*, 6, 9, 11; N. Elizabeth Schlatter, *Museum Careers: A Practical Guide for Students and Novices* (Abingdon: Routledge, 2016).
90. IWM, MA, EN1/1/REP/032, see reports therein.
91. Schlatter, *Museum Careers.*
92. Child, *Organization*, 7, 12.
93. Ibid., 8, 12–13; Philip W. Deans, 'Curating the Great War at the Imperial War Museum during the Second World War', in *Curating the Great War*, ed Paul Cornish and Nicholas J. Saunders (Abingdon: Routledge, 2022), 86–87.
94. Genoways and Ireland, *Museum Administration*, 40.
95. Child, *Organization*, 8, 13.
96. Miller, *Anatomy*, 15.
97. Richard A. Chapman, *Civil Service Commission, 1855–1991: A Bureau Biography* (London: Routledge, 2004), 41–45.
98. IWM, MA, EN1/1/REP/032, see reports therein.
99. IWM, EN1/1/STA/004, see documents therein.

100. IWM, MA, EN1/1/REP/032, see reports therein.
101. Ibid.
102. Timothy Ambrose and Crispin Paine, *Museum Basics: The International Handbook*, 4th edn (Routledge: Abingdon, 2018), 370–374.
103. Esmé, Countess of Carlisle, *A History of the Commission* (London: Museums and Galleries Commission, 1991); Geoffrey Lewis, *For Instruction and Recreation: A Centenary History of the Museums Association* (London: Quiller, 1989); IWM, MA, EN1/1/MUSG/013, see documents therein; IWM, MA, EN1/1/MUSG/011, see documents therein; IWM, MA, EN1/1/MUSG/021, see documents therein.
104. IWM, MA, EN2/1/COB/040, see documents therein; IWM, MA, EN2/1/COB/049, see documents therein.
105. See Caroline Shenton, *National Treasures: Saving the Nation's Art in World War II*, pbk edn (London: John Murray, 2022), 88–89.
106. IWM, MA, EN1/1/REP/007, '7th Annual Report (4th Report of the Board of Trustees), 1923–1924', 1.
107. IWM, MA, EN2/1/STA/009/4, anon. to Stokes, 11 January 1945, 1.
108. *Imperial War Museum Act*, 1920.
109. Ibid.
110. Ibid.
111. Child, *Organization*, 8.
112. Roger Smither and David Walsh, 'Unknown Pioneer: Edward Foxen Cooper and the Imperial War Museum Film Archive, 1919–1934', *Film History* 12 (2000): 187–203.
113. Noble Frankland, *History at War: The Campaigns of an Historian* (London: Giles de la Mare, 1998), 165.
114. IWM, MA, EN1/1/MUS/024, see documents therein.
115. Eilean Hooper-Greenhill, *Museums and their Visitors* (London: Routledge, 1994), 20.
116. IWM, MA, EN1/1/MUS/024, see documents therein.
117. Bruce A. Campbell and Christian Baars, 'Introduction', in *The Curation and Care of Museum Collections*, ed Bruce A. Campbell and Christian Baars (Abingdon: Routledge, 2019), xvi.
118. Gail Dexter Lord, 'The Exhibition Planning Process', in *The Manual of Museum Exhibitions*, ed Barry Lord and Gail Dexter Lord (Walnut Creek, CA: AltaMira, 2001), 1.
119. Suzanne Keene, *Fragments of the World: Uses of Museum Collections* (Oxford: Elsevier Butterworth-Heinemann, 2005), 26–31.
120. Ibid., 31.
121. Ibid., 27.
122. Ibid., 27–29.
123. Ibid., 29.

124. Ibid., 29–31.
125. See newspaper art supplements, the *Museums Journal* and tourist websites for examples.
126. David Dean, *Museum Exhibition: Theory and Practice*, pbk edn (London: Routledge, 1996), 2.
127. Flora E. S. Kaplan, 'Exhibitions as Communicative Media', in *Museum, Media, Message*, ed Eilean Hooper-Greenhill, pbk edn (Abingdon: Routledge, 1999), 37–58.
128. Doug Ashford, 'The Exhibition as an Artistic Medium', *Art Journal* 57 (1998): 28–37.
129. Elizabeth Bogle, *Museum Exhibition: Planning and Design* (Lanham, MD: AltaMira, 2013).
130. Lewis, *Instruction and Recreation*, 1–57.
131. Henry Miers, *A Report on the Public Museums of the British Isles (Other Than the National Museums)* (Edinburgh: Carnegie United Kingdom Trustees, 1928); Royal Commission on National Museums and Galleries, *Interim Report* (London: HMSO, 1928); Royal Commission on National Museums and Galleries, *Final Report, Part I: General Conclusions and Recommendations* (London: HMSO, 1929); Royal Commission on National Museums and Galleries, *Final Report, Part II: Conclusions and Recommendations Relating To Individual Institutions* (London: HMSO, 1930); S. F. Markham, *A Report on the Museums and Art Galleries of the British Isles (Other Than the National Museums)* (Edinburgh: Carnegie United Kingdom Trustees, 1938).
132. Catherine Pearson, *Museums in the Second World War: Curators, Culture and Change*, ed Suzanne Keene (Abingdon: Routledge, 2017), 29–30, 39, 45.
133. Miers, *Report*, 31.
134. Markham, *Report*, 8–9.
135. Ibid., 9.
136. IWM, MA, EN1/1/REP/004, 'Report of the Imperial War Museum, 1920–1921'; IWM, MA, EN1/1/REP/005, 'Fifth Report of the Imperial War Museum (Second Report of the Board of Trustees), 1921–1922'; IWM, MA, EN1/1/REP/006, '6th Annual Report (3rd Report of the Board of Trustees), 1922–1923'; IWM, MA, EN1/1/REP/007, '7th Annual Report (4th Report of the Board of Trustees), 1923–1924'; IWM, MA, EN1/1/REP/008, '8th Annual Report (5th Report of the Board of Trustees), 1924–1925'; IWM, MA, EN1/1/REP/009, '9th Annual Report, 1925–1926'; IWM, MA, EN1/1/REP/010, '10th Annual Report, 1926–1927'; IWM, MA, EN1/1/REP/011, '11th Annual Report, 1927–1928'; IWM, MA, EN1/1/REP/012, '12th Annual Report, 1928–1929'; IWM, MA, EN1/1/

REP/013, '13th Annual Report, 1929–1930'; IWM, MA, EN1/1/ REP/015, '14th Annual Report, 1930–1931'; IWM, MA, EN1/1/ REP/032, see reports therein; 'News from the Museums', *Museums Journal* 36 (1936): 263–265; 'News from the Museums', *Museums Journal* 37 (1938): 521–526; 'News from the Museums', *Museums Journal* 38 (1938): 199–204.

137. Owain Rhys, *Contemporary Collecting: Theory and Practice* (Edinburgh: MuseumsEtc, 2011).
138. John Tosh, *The Pursuit of History: Aims Methods and New Directions in the Study of Modern History*, rev. 3rd edn (Harlow: Longman, 2002), 54–82; Nikki Sullivan, 'Contemporary Collecting in a History Museum', MuseumNext, https://www.museumnext.com/article/museums-contemporary-collecting/ (18 March 2024); Leonie Hannan and Sarah Longair, *History Through Material Culture* (Manchester: Manchester University Press, 2017).
139. Owain Rhys and Zelda Baveystock, 'Introduction', in *Collecting the Contemporary: A Handbook for Social History Museums*, ed Owain Rhys and Zelda Baveystock (Edinburgh: MuseumsEtc, 2014), 15.
140. Cornish, '"Sacred Relics"', 37–40.
141. Jane C. Milosch and Nick Pearce, eds, *Collecting and Provenance: A Multidisciplinary Approach* (Lanham, MD: Rowman and Littlefield, 2019).
142. Gaynor Kavanagh, 'History in Museums in Britain: A Brief Survey of Trends and Ideas', in *Social History in Museums: A Handbook for Professionals*, ed David Fleming, Crispin Paine and John G. Rhodes (London: HMSO, 1993), 16; Gaynor Kavanagh, 'Museum as Memorial: The Origins of the Imperial War Museum', *Journal of Contemporary History* 23 (1988): 94.
143. Quoted in Cornish, '"Sacred Relics"', 37.
144. Franz Uhle-Wettler, 'War', in *Brassey's Encyclopedia of Military History and Biography*, ed Franklin D. Margiotta (Washington, DC: Brassey's, 1994), 1047.
145. Cornish, '"Sacred Relics"', 37.
146. Kavanagh, *Museums*, 126–127.
147. Ibid., 131.
148. Paul Cornish, 'Extremes of Collecting at the Imperial War Museum, 1917–2009: Struggles with the Large and the Ephemeral', in *Extreme Collecting: Challenging Practices for 21st Century Museums*, ed Graeme Were and J. C. H. King (New York, NY: Berghahn, 2012), 158; Kavanagh, 'History in Museums', 20.
149. Malvern, 'War', 188.
150. Cornish, '"Sacred Relics"', 37; see Susan M. Pearce, *Museums, Objects and Collections: A Cultural History* (London: Leicester University Press, 1992), 84–88.

151. Wellington, *Exhibiting War*, 108.
152. Sue Malvern, *Modern Art, Britain and the Great War: Witnessing, Testimony and Remembrance* (New Haven: Yale University Press, 2004), 14.
153. Kavanagh, *Museums*, 138.
154. Wellington, *Exhibiting War*, 114.
155. Kavanagh, *Museums*, 139.
156. Ibid.
157. Malvern, *Modern Art*, 71.
158. James Fox, *British Art and the First World War, 1914–1924* (Cambridge: Cambridge University Press, 2015), 134.
159. Malvern, 'War', 188.
160. Alys Cundy, 'War, Memory and the Senses in the Imperial War Museum, London', in *Modern Conflict and the Senses*, ed Nicholas J. Saunders and Paul Cornish (Abingdon: Routledge, 2017), 363–364.
161. Cundy, 'Thresholds of Memory', 253.
162. Malvern, 'War', 188–189.
163. Ibid., 188; Paul Cornish, 'Introduction', in *Curating the Great War*, ed Paul Cornish and Nicholas J. Saunders (Abingdon: Routledge, 2022), 2–3.
164. Condell, 'History and Role', 31.
165. Terry Charman, '"A Museum of Man's Greatest Lunatic Folly": The Imperial War Museum and its Commemoration of the Great War, 1917–2008', in *A Part of History: Aspects of the British Experience of the Western Front* (London: Continuum, 2008), 104; IWM, MA, EN1/1/MINI/002, see documents therein; IWM, MA, EN2/1/STA/009, see documents therein.
166. Cornish, '"Sacred Relics"', 37–40.
167. Aardman, *Flight of the Stories—Our New First World War Galleries*, YouTube video, IWM, https://www.youtube.com/watch?v=-lZUTMS8uLA (18 March 2024).
168. Cundy, 'Thresholds of Memory', 254–255.
169. Ibid.
170. Smither and Walsh, 'Unknown Pioneer', 193–194.
171. Cornish, '"Sacred Relics"', 46.
172. Bogle, *Museum Exhibition*.
173. H. Foster, 'The Imperial War Museum: Its New Home at Southwark', *Museums Journal* 36 (1936): 220.
174. IWM, MA, EN1/1/REP/032, see reports therein for 1935–1939.
175. IWM, MA, EN1/1/REP/032, '21st Annual Report of the Director-General to the Board of Trustees, 1938–1939', 4.

176. Ibid., 6; James Taylor, '*Your Country Needs You*': *The Secret History of the Propaganda Poster* (Glasgow: Saraband, 2013), 11–12.
177. Cornish, 'Introduction', 2–3.
178. Joyce Apsel, *Introducing Peace Museums* (Abingdon: Routledge, 2016), 12.
179. James Wallis and James Taylor, 'The Art of War Display: The Imperial War Museum's First World War Galleries, 2014', in *Commemorative Spaces of the First World War: Historical Geographies at the Centenary*, ed James Wallis and David C. Harvey (Abingdon: Routledge, 2018), 102.
180. John A. Byrne, 'The Virtual Corporation', *International Business Week*, 8 February 1993.

CHAPTER 4

Continuing a Civic Service, 1939–1945

Introduction

Having analysed the Imperial War Museum as an institution, organisation and museum during Chap. 3, this study now examines three crisis-conducive situations foregrounded in Chap. 1 which the institution encountered over the Second World War era. The first, addressed by the current chapter, was a crisis of inactivity, arising from disruption to its civic service provision. As explained in Chap. 2, crises are highly obstructive phenomena. They pose challenges for museums, which cause short-, medium- or long-term disruption. Oftentimes, such situations will require recourse to resilience, formally introduced in Chap. 2, when being overcome. Through exploring this crisis-conducive situation, the current chapter expands on the concept of resilience. In doing so, it shows that organisational resilience underlies all successful museum responses to challenging situations, including crises, meaning this quality must be developed for when difficult times occur.

The historical context of the current chapter is the Second World War years themselves, 1939-1945, and the general ramifications which the war posed work and everyday life in the United Kingdom. At this time, the need to fight and win the war, while protecting people from its effects, saw national and local government create laws and policies which tightly controlled what organisations, communities and individuals could, should and must do, and *vice versa*.[1] Even in situations outside regulatory control,

P. W. Deans, *Crisis, Reinvention and Resilience in Museums*,
https://doi.org/10.1007/978-3-031-67806-6_4

the conditions created by the conflict meant that everyday life did not continue unaffected. Dangers arising from the violent and destructive nature of the war, plus the general scarcity of resources available, meant organisations, communities and individuals became unable to follow normal patterns of behaviour.[2] Museums were no exception. Controls on visitor admission, the forced closure of premises and the requisition of facilities meant that, in many cases, the normal operations of museums could not continue while the country remained a war zone.[3] This was the case for the Imperial War Museum, which ended up being prevented entirely from dispensing its usual civic services. As a result, the museum had to diversify and work creatively in order to continue to justify its existence and to remain immediately worthy of societal support.

The current chapter draws on two concepts in analysing the crisis of inactivity at the Imperial War Museum. One is 'civic service', representing museum work that supports national and local civil society. The other is 'resilience', previously introduced in Chap. 2, which represents the capacity of organisations to withstand environmental turbulence.[4]

The current chapter addresses the above over five substantive sections. The first two introduce and explicate the concepts of civic service and resilience respectively, while the third considers how museum resilience depends on an ability to perform a meaningful civic service. Both sections elucidate their respective concept/s and any relevant issues, ideas and activities. The main section thereafter then profiles ways that the Second World War disrupted normal civic services at the Imperial War Museum delivered routinely before the conflict, as well as methods employed by the museum to overcome this disruption. It starts by focussing on public-facing aspects, and then latterly focusses on non-public-facing aspects.

Civic Service

The concept of 'civic service' requires some elucidation. This is because it can mean different things to different people at different times, depending on their views over what the activity entails and the situation being confronted.[5] Some people understand civic service as a private, informal and voluntary initiative addressing specific need. Others, however, understand it as a public, formal and institutionalised initiative towards attaining instrumental value.[6] The concept can also be understood as a derivative of, or alternative to, national or military service, and discussed *apropos* people or organisations.[7] This explains why Richard Danzig and Peter Szanton consider civic service an entirely arbitrary idea.[8] It also perhaps explains why, in the British museum context, Peter Latchford perceives 'no formal UK civic museum category or definition'.[9] Thus, over the current chapter, a simple yet effective understanding of civic service is applied to analyse the Imperial War Museum.

'Civic' has two meanings in British English. It primarily pertains to civil society. This derives from the Latin word *civitas*, roughly meaning the 'citizenry of a state'.[10] But, as Chris Brink points out, civic also pertains to towns, cities or other local geopolitical areas.[11] The 'civic museum', therefore, may mean two things. In the former sense, it could pertain to a museum which addresses the wants, needs and interests of civil society. In the latter, the museum of the town, city or other geopolitical area wherein the museum is located. By synthesising these ideas, an insightful definition presents itself for describing a museum's civic service. This is any practicable service that supports the community big or small in which the museum resides by facilitating or positively enhancing the lives of its constituents.

It would be impossible to analyse every instance of civic service performed by the Imperial War Museum during the Second World War and avoid treating them superficially or repetitively. There are just too many instances recorded in the museum archive. Consequently, several of the more impactful examples have been selected for analysis, either alone or grouped under a theme. These include work by the museum to, firstly, produce an exhibition for the general public during the 'Bore War' period—more commonly known by the American name 'Phoney War'—in 1939-1940, secondly, interact with the public after sustained air raids on London forced its closure and, thirdly, support the British war effort.

Resilience

The current chapter now returns to the concept of organisational resilience. Surprisingly, this issue has seldom featured in museum literature, hitherto. On the few occasions it does, the discussion usually explores how museums can develop resilient qualities, or more rarely how museum resilience can be analysed and assessed.[12] As introduced in Chap. 2, organisational resilience involves the ability to restore from, withstand or adapt to difficult or changing conditions while continuing to maintain operationality.[13] It is a quality facilitated by proactive actions and behaviours including embracing new aims and objectives, watching out for new or emerging opportunities and adopting new capabilities.[14] Resilience lies at the heart of museum longevity. Indeed, a museum without resilience will unlikely exist for long, as they will lack the wherewithal to manage difficult situations. Hence, although conceptually independent, many ideas, attributes and features of resilience feed into crisis management.

Two contrasting conceptualisations of resilience are forwarded by Keith Shaw and Louise Maythorne. Drawing on ideas from Bridgit Maguire and Sophie Cartwright, they elucidate the first as a 'recovery' process, where demonstrable resilience involves persevering through turbulence.[15] This links with the 'bounce back' and 'ability to absorb' understandings of resilience introduced in Chap. 2.[16] The more unencumbered the museum remains by/after turbulence, the greater its resilience.[17] The second conceptualisation is elucidated as a 'transformation' process, where demonstrable resilience involves adapting away from turbulence.[18] This links with the 'positive adaptability' understanding of resilience also introduced in Chap. 2.[19] Here, the museum makes requisite changes to thrive in the new context, not just survive.[20] Mike Raco and Emma Street frame this notion as a 'radical' understanding of resilience which rejects the idea that an organisation suddenly impacted by turbulence might previously have been resilient.[21] For example, Richard J. T. Klein, Robert J. Nicholls and Frank Thomalla posit: '[I]f a megacity is struck by a disaster, it follows that the original state was one in which it was vulnerable to the disaster in the first place'.[22] The viability of such an interpretation is debatable. No museum can ever confidently claim infallibility to a threat which has not been conceived. Moreover, the effort of striving for infallibility will always fail and could prove organisationally crippling. It is, however, helpful in serving to remind that no museum can enter a period of stasis and survive indefinitely. As discussed in Chap. 2, museums must continually update themselves through making value adjustments against an evolving society to remain relevant and functional.

Museums develop resilience in various ways. Mara Cerquetti and Eleonora Cutrini raise three 'enabling dimensions' to museum resilience: 'structures', 'people' and 'relationships'.[23] The first and most commonly studied dimension in organisational studies is the building and maintaining of strong physical or semi-physical assets.[24] Without robust facilities and systems, museums cannot easily operate in calm periods, let alone turbulent ones. But physical durability alone should not be relied on during adversity. The most effective resilience draws on a full spectrum of different organisational strengths, including intangible attributes.[25] This is where the other two above dimensions come into play. The second concerns the recruitment of competent agents.[26] Staff who are knowledgeable about organisational systems, schedules, and routines; who are proactive, resourceful and creative; and who possess troubleshooting and problem solving skills: such personnel can go far in building up a museum's resilience.[27] This is particularly the case if their attributes have been identified and nurtured by the museum which employs them.[28] And the third dimension involves establishing collaborative networks and relationships.[29] These enable museums to draw on wider pools of physical and intellectual resources towards maintaining and restoring operations. While preventing and recovering from disaster is an obvious example, it also encompasses working in partnership to reach and serve wider and more diverse audiences.[30]

Resilience Through Civic Service

Marrying up the above issues and ideas on civic service provision and resilience in the museum context, Robert R. Janes makes the argument that delivering a civic service is essential to building museum resilience.[31] Quoting from Stephen Weil, he buttresses this argument by asserting: 'Survival [...] is not [the museum's] purpose'.[32] His implication here is that museums cannot exist for the sake of existing. Rather, they can only exist by serving one or more civic communities.[33] To Janes, museum resilience relies on museums maintaining a civic service which meets the needs of society. Cerquetti and Cutrini concur, arguing that 'if a museum fosters resilience culture and community empowerment, local stakeholders are more aware of the[ir] value [...], ensuring its survival for future generations'.[34] Without some manifesto and programme for a public good, their point may come into question and their legitimacy and all this entails put in doubt.[35]

Concern for this is demonstrated (if not explicitly so) by the influential 1938 Markham Report introduced in Chap. 3. The review made it clear that the point of museums was to serve the public, whatever an individual user's context. It also argued that museums, wherever they are and whatever their remit, should proactively address the cultural inequalities pervading society.[36] This finding is significant by setting the civic service performed at the Imperial War Museum during the Second World War era against a backdrop of broader museum sector recognition about the general need to serve society.

Resolving a Crisis of Inactivity at the Imperial War Museum

Producing a Public Exhibition

At the onset of the Second World War, the British government was concerned about aerial attacks on urban population centres. This prompted it to enforce a blackout that caused the closure of many public gathering places.[37] But despite fears to the contrary, no bombs fell immediately. As a result, people started calling the war the 'Bore War' in these early months, owing to the fighting's perceived distance from the British Isles and the limited recreational facilities now available.[38] The government revised this decision however following mounting protests including lobbying by the Museums Association and the Standing Commission on Museums and Galleries, allowing entertainment and cultural venues to reopen.[39]

The first national museum in London to readmit the public was the Victoria and Albert Museum. Hasty preparations following the government's U-turn saw several gallery spaces reopen there on 11 January 1940.[40] In response, the Trustees of the Imperial War Museum instructed Leslie Bradley, its Director-General, to do the same and reinstate some public-facing civic service at the museum.[41] This instruction points towards a pursuit of 'business continuity': the process and discipline of 'avoid[ing] any interruption that could lead to either significant losses or failure to achieve the organisations principle objectives'.[42] Business continuity supports any work involved in developing museum resilience. If organisational resilience generally involves the ability to restore from, withstand or adapt to difficult or changing conditions while continuing to maintain operationality, business continuity is the underpinning work of maintaining critical museum activities.[43] For the Imperial War Museum, ensuring business continuity at this time saw it curate an exhibition on the museum's three ground floor galleries which could be accessed by the public at large.

The unnamed exhibition arising from this decision opened on 29 January 1940 and enabled the Imperial War Museum to re-establish a public-facing civic service. Curated mostly using longstanding material from the collection that had not hitherto been evacuated, it included models, artefacts, artworks and documents displayed over the three main ground floor galleries: the Naval Gallery, the Army Gallery and the Picture

Gallery. In the Naval Gallery, visitors could see models of ships and boats, the actual Short Seaplane which flew at the Battle of Jutland, and specimens of weaponry such as torpedoes, mines and depth charges. Alongside these exhibits was material pertaining to women's war work.[44] In the Army Gallery, visitors could see artillery, camouflage and the personal equipment of soldiers. Alongside these were various other items representing contraband, espionage and counter-espionage, diplomatic and military documentation, propaganda, 'and a diversity of other objects'.[45] And in the Picture Gallery, visitors could see lithographs and drawings which conveyed anti-German sentiment. Items from the Air Gallery situated on the floor above were also sporadically displayed across all three of these ground floor galleries.[46]

Each display comprised material relating almost exclusively to the First World War. There existed two exceptions however, perhaps reflecting the Trustees' stated hope that 'some of the more popular new material collected by the Departments might also be [...] shown'.[47] One was a model representing HMS *Ajax* which had shot to public attention the previous month following its part in the Battle of the River Plate—the Allies' first naval victory of the war.[48] The second was the infamous document—the Joint Declaration between the United Kingdom and Germany—bearing Neville Chamberlain and Adolf Hitler's signatures, which Chamberlain had proudly brandished at journalists on exiting the aeroplane following his return from negotiations with Hitler and the Munich Agreement.[49] The latter became an item of considerable public interest, with the *Daily Mail* and the *Daily Mirror* both publishing short articles solely notifying readers about its display.[50] Mollie Panter-Downes, a novelist and writer for the *New Yorker*, commented in an article following her visit to the exhibition about the sense of irony she felt by the document's presence.[51]

Although very little literature and no photographs from the exhibition are seemingly extant, the *Museums Journal* and *Times* offer some useful insights into it, particularly in regard to how the First World War material on display made vivid connections with the contemporary.[52] The following comprise just three such examples. In the Naval Gallery, photographs depicted Hermann Göring—who, during the Second World War, held many senior political and military positions in the Nazi state—as commander of his German fighter wing during the First World War.[53] In the Army Gallery, gasmasks sat alongside other personal protective equipment from the First World War.[54] And in the Picture Gallery, a drawing by Henry Rushbury depicting Winston Churchill addressing an audience at

Central Hall, Westminster, on 4 July 1918, was displayed with a quotation from his speech inscribed underneath. It stated: 'Germany must be beaten, must know she is beaten, must feel she is beaten!'[55] The contemporaneous nature of this last exhibit is extraordinary. By February 1940, Churchill was not yet Prime Minister. He had however returned to government as First Lord of the Admiralty, an office which he also briefly held during the First World War. Churchill was an arch critic of Chamberlain's policy of appeasement towards Nazi territorial ambitions, and often warned against treating Germany as harmless in a spirit that evoked the sentiment represented by Rushbury's drawing.[56] As the *Illustrated London News* commented during its article on the exhibition, the Churchill represented in the drawing bore an 'almost identical' resemblance to the Churchill who just one week before the article's publication had spoken in Manchester about the need to fight and to win this new war.[57]

In utilising the material and spaces at their disposal, the staff performed what has become known amongst organisational resilience theorists as 'bricolage': operating modes that accommodate 'the capacity to improvise and apply creativity in problem-solving'[58]—a reuse of the term originally coined by Claude Lévi-Strauss in *The Savage Mind*.[59] This activity might sound counterintuitive during difficult situations such as crises. It is, however, important in facilitating organisational resilience.[60] Bricolage engages an agent's skill set outwith the constraints placed on them by organisational convention, enabling a reapplication of existing organisational knowledge and practices into novel combinations and configurations. These go on to form informal and temporary organisational structures, which support the museum through turbulence emanating from the social system until normal operations can be resumed. In the case of the Imperial War Museum, the process involved its staff synthesising knowledge and experience of previous exhibition practice with knowledge of the resources available to create a temporary exhibition which met the requirements of the museum while the Second World War inhibited usual public programming.

Despite focussing on the First World War, this exhibition was about the Second World War. As the article from *Museums Journal*, written by Ernest Blaikley, the art curator of the Imperial War Museum, explained, 'the result is an extraordinarily vivid display which illustrates, at almost every point, the present wartime life of the community'.[61] The *Times* article similarly explained how 'exhibits with a particular bearing on the present-day war conditions have been selected for display'.[62] The upshot of this was an exhibition with a distinctly contemporary relevancy about it. In the museum context, relevancy involves identifying external challenges and making positive differences against them.[63] The Imperial War Museum did this by guiding visitors towards 'meaning-making'—the process whereby display material, such as objects, artwork or text, attains meaning to individual visitors through syntheses of deliberate curatorial decisions and individual 'visitor contexts': the prevailing personal, socio-cultural and physical circumstances accompanying them—which, firstly, encouraged the formation of opinions legitimising Britain's participation and, secondly, provided visitors with psychological nourishment.[64] In doing so, it maintained a civic service through the exhibition as defined above, further demonstrating the museum's business continuity. It also illustrates the phenomenon conceived by David Farrell-Banks as a 'moving moment'. These are representations of past events which, organically or purposely, have transitioned between different physical, conceptual, intellectual and media spaces and domains; developed to resonate with a particular present context; elicited an emotional response and, through the combined effect of the previous three, roused political action.[65]

The opinion forming, or propagandist, aspect to the exhibition's meaning-making derived from its curatorial framing of Germany: a hawkish, hostile country, which viewed opponents with contempt. David Welch describes 'propaganda' as, essentially, 'the deliberate attempt to influence public opinion through the transmission of ideas and values for a specific purpose'.[66] That the exhibition tried to stoke anti-German sentiment is apparent from not only the collections on display and the way Blaikley writes about them for his article in the *Museums Journal*, shown above, but also the way the Imperial War Museum apparently negatively evolved its interpretation of Germany following the war's onset. The museum's peacetime policy until the Second World War had been to treat any issues involving Germany sensitively and un-antagonistically.[67] For example, in 1933, Bradley found himself placating a distressed German visitor who had taken offence at some interpretation. 'We have endeavoured to avoid giving offence to our former enemies',[68] he explained by letter. But the declaration of war changed this policy. In interpretation accompanying the Rushbury drawing, Panter-Downes recalled reading that 'until recently delicacy has prevented them from showing the drawing, but that they are doing so now because "the sentiments expressed in Mr. Churchill's speech have once more become appropriate and sensible"'.[69]

The psychological nourishment aspect by contrast derived from the ability of the exhibition to contextualise and interpret the new war to the public and requires deeper explanation. Over 1938-1940, the British population experienced what Julie Gottlieb has termed a 'war of nerves'. This concept refers to the condition of society at the time as documented in Mass Observation records, characterised by nervous exhaustion from the worry and fatigue over the deteriorating European situation and the threats posed by another war with Germany.[70] The exhibition offered a way for frightened individuals to come to terms with the national situation and their position therein—fear being not an uncommon feeling amongst Londoners at this time.[71] It provided people an opportunity to reach an understanding. This assertion draws on psychological theory concerning anxiety. Seeking knowledge about threats shrouded in ambiguity has long been observed to help reduce nervousness and unease caused by uncertainty.[72] George Loewenstein's argument for this is that inquisitiveness arises 'when attention becomes focussed on a gap in one's knowledge. Such information gaps produce the feeling of deprivation labelled *curiosity*. The curious individual is motivated to obtain the missing information to reduce or eliminate the feeling of deprivation'.[73] Obviously, feelings of deprivation can take many subjective forms. One example is the feeling which may arise when individuals find themselves deprived of enough information to maintain the feeling of agency over their lives: a lack of perceived control being an underlying cause of anxiety.[74] With that in mind, the exhibition can be viewed as having offered nourishment for the public psychological condition through interpreting the new war, thereby nourishing the public's curiosity and helping alleviate its war of nerves.

The Imperial War Museum demonstrates qualities of both the 'recovery' and 'transformation' conceptualisations of resilience through this example of it reinstating some public-facing civic service.[75] On one hand, the work represented a development in prevailing exhibition practice there by addressing issues and ideas not previously engaged with. But on another, it represented a continuation of existing practice through drawing on pre-existing display material and design.[76] Thus, the exhibition can be understood as a redisplay: new, yet familiar at the same time.[77] It did not represent a 'revolution' or 'reinvention' of exhibition practice, for these terms imply radical change.[78] Rather, it represented ideas around resilience forwarded by Janes, where some museum moves to accommodate the turbulence emanating from the social system, but without wholly changing their *raison d'être* and rationale.[79]

The public wartime exhibition lasted just over eight months until 10 September when the Blitz forced the Imperial War Museum to close again. This time, it remained closed for the remainder of the war in Europe.[80] The exhibition had proved popular, with 65,496 visitors attending in total.[81] This was despite its short lifespan and the restrictions imposed on public access by the Metropolitan Police and London Borough of Southwark: that no more than 200 individuals could attend at any one time, that access was restricted to daylight hours only and that, initially, the museum remained closed on Sunday.[82] Later, visitor figures for Easter Sunday state that 1200 people attended in just four hours, such was the public interest.[83]

Maintaining Interaction with Constituents

The Imperial War Museum was forced to close again on 10 September 1940 after a large bomb fell in the near vicinity without detonating. It would not reopen again until after the European conflict had ended; the intensity and unpredictability of the Blitz, plus the limited local shelter capacity, rendered mass public admittance too unsafe.[84] This posed a problem to the museum, as regular public interaction and audience/customer development is essential for any organisation.[85] Without the legitimacy that comes from continual use by the public, museums lose the support they require.[86] The closure of the Imperial War Museum to mass public admittance did not stop it from interacting with the public, however. On the contrary, interactions with the public continued via alternative, if more constrained, methods.

The first of two ways that the Imperial War Museum continued interacting with the general public during the Second World War was through written correspondence. While it had regularly dealt with letters from the public before the war, those answered during the war assumed greater importance. They ensured the museum remained connected with users. Fundamentally, organisational resilience is the ability to continue delivering a 'required capability' in the face of turbulence: 'capability' being what Timothy L. J. Ferris et al. consider 'the ability to achieve a specific objective under stated conditions'.[87] For the Imperial War Museum, this comprised its civic service, something which hitherto had been delivered mainly through public-facing activities. Now however, the capability needed to be delivered differently. That staff could immediately fall back on written communication offers a further demonstration of acquired resilience.[88] Indeed, correspondence preserved in the museum archive suggests the Imperial War Museum received heterogeneous letters. Each example includes a response signed by the Director-General, Bradley.[89]

Most letters sent by the public to the Imperial War Museum during the Second World War simply enquired whether it was open or not. A significant minority however concerned other matters. Some of these were quite irrelevant to the museum. On 28 March 1940, a man wrote 'to ask you if you could tell me the value of a tin box of 1 pound of chocolate which was given to my father by Queen Victoria at Christmas time in 1900'.[90] Others were more relevant. On 15 June 1943, a soldier in the British Army wrote requesting information about the decorations worn by King George VI when wearing military uniform 'in order to assist in a large coloured photo of H.M. which I am colouring'.[91] And a few comprised pleas for assistance. On 17 October 1944, a woman wrote after 'a nephew of a friend', who had been a paratrooper on D-Day, visited her 'with [...] several souvenirs + amongst them [...] a map which he had captured from a German [...] of the Southern Coast of England'.[92] This map reportedly became lost while in her possession. She therefore wrote asking if anyone at the museum may 'have come in touch with, or are likely to come into touch with anyone owning a similar map, who would be prepared to sell it to me?'[93] Each of these examples received a personal response from Bradley, offering information and support wherever he could.[94]

The second way that the Imperial War Museum continued interacting with its constituents was through offering private admittance, typically for the purposes of conducting research. That is, pre-arranged, limited and supervised access by private persons. Of course, as a public institution, such restricted admittance policies would not have been routine. But the conditions created by the Second World War were far from regular, therefore requiring stopgap measures to realise the museum's required capability. This phenomenon is known as 'scaffolding': where temporary, flexible, generative and constitutive structures metaphorically buttress some initiative or activity.[95] Andy Clark helpfully elucidates this by describing such manifestations as 'augmentations that allow us to achieve some goal that would otherwise be beyond us'.[96] Since the purpose of resilience is the continuation of organisational operations, it follows that scaffolding can be a way of ensuring this outcome. Where unable to achieve required capability, museums can scaffold operations by adding or amending organisational components as they are required.[97] Indeed, through instituting another mode of interaction with the public, the Imperial War Museum not only demonstrated pre-existing resilience, but also an ability to generate extra sources of resilience.

Instances of private admission to the Imperial War Museum during the Second World War usually resulted from enquiries about a research question. The unpublished 'War History of the Imperial War Museum' details various such occasions, particularly during 1939-1940, where individuals attended for research. It remarks that there existed several main classes of enquiry. One was new service personnel waiting to join their ship, regiment or squadron, who sought understanding from historical accounts about what the new war might entail for them. Another was journalists who sort either photographs depicting the previous war which they could use to illustrate their articles, or information on the war work or service performed during the previous war by individuals now in authority.[98] One such example where the Imperial War Museum admitted a member of the public was the above-mentioned soldier who sought information about The King's military decorations. On receiving this request, Bradley contacted Buckingham Palace for the information. The palace responded by sending a coloured drawing depicting The King in full uniform.[99] From this, Bradley produced facsimiles of the decorations, which he invited the soldier to view.[100]

Providing Research Facilities

In addition to interacting with the public via correspondence or private admission, the Imperial War Museum maintained core business continuity by actively delivering various non-public-facing services.[101] This further ensured that its required capability of delivering a civic service could be continued.[102] Such services, considered below, were conceivably more alien to the museum, in philosophy if not practice, than those already discussed. Their existences however can be explained by the fact that the United Kingdom was again, as over 1914-1918, fighting a total war.[103] Accordingly, communities and institutions throughout society had to adapt to exist in a way that helped the war effort. Museums were no exception. Indeed, the Standing Commission on Museum and Galleries reported after the war in 1948 how many performed a heterogeneous array of civic services that in peace time would have been inconceivable, but which total war rendered logical and necessary for their continued legitimacy.[104] Museums big and small would also have been expected to play their part.

This situation resonates with Janes' ideas about museum resilience discussed above. He argues for museums to make proactive, positive interventions in society which build up their attributes of resilience[105]—a view resonating with the adage invoked recently by Sharon Heal from the Museums Association: 'if you are not part of the solution, you are part of the problem'.[106] Implicit here is an assumption that museums are only as secure as they are both proactive and, more importantly, needed by society.[107] Granted, in making this argument, Janes does not consider a context of total war, but rather human-induced climate change. Nevertheless, it translates to this case. During the Second World War, if museums were not actively supporting the war effort, they could be seen as hindering it, risking negative legal, political or cultural consequences. Museum resilience therefore also depends on continuing a civic service, whatever the service may be. Whether it conforms to the museum's typical or commonly accepted paradigm or not is of secondary importance.

One of the more conventional non-public-facing civic services which follow neatly on from the previous public-facing example was by facilitating war research. During the Second World War, many individuals acting for the government visited the Imperial War Museum to undertake war research. Its photographs were of particular interest in this regard. They could inform officials on many and varied issues. Subjects which underwent investigation there included, but were not limited to, advising on the erection of camouflage for concealing strategic assets; the design of steel helmets for civilian issue; the development of gas protection for buildings, people and animals; and the construction of trenches, shelters and dugouts.[108] This work can also be seen as bricolage discussed above.[109] Before and after the Second World War, the museum had adopted a style akin to an early peace museum.[110] But once the United Kingdom was formally at war with Germany, this stance quickly became untenable. Operational requirement meant that the museum needed to temporally accept and institute new positions and policies over the use and rationale of its collection.

A remarkably detailed collection of Home Office papers, located at The National Archives, provides a particularly vivid illustration of the use of the material held by the Imperial War Museum in this regard. On 24 November 1939, the museum received a visit by Professor J. D. Bernal and his assistant to undertake government war research. Bernal was an Irish physicist who had adjourned an academic career and taken up employment with the Department of Home Security at the Research and Experiments

Department in Princes Risborough.[111] The reason for the pair's visit was to assess the destructiveness of aerial bombs on ground targets. A hand-written report produced after the visit details all the material that they consulted there and its informative value.[112] Of these, photographs and uniquely held London Fire Brigade reports were considered most beneficial.[113]

Loaning Objects to Organisations and Initiatives

A second non-public-facing civic service which the Imperial War Museum performed during the Second World War involved loaning out items from its collections for use towards war initiatives arranged by other organisations. These typically comprised curated exhibitions concerning local and/or national issues and causes, such as the London *Gas in War* and *Fuel Economy* exhibitions of 1941 and 1942 respectively, and the regional and national exhibitions about the war funding scheme National Savings.[114] Each act of support required the museum to draw on bricolage for going against its peacetime philosophy.[115]

Returning Objects to Their Original Functionality

That the Imperial War Museum allowed collection items to be used towards the country's war effort clearly contradicted its pre-Second World War anti-war mission. But in the main, the basic underlying activities of loaning out objects for research and display were not uncharacteristic, as attested by the Markham Report. Accordingly, from a practical perspective, most would have been easily implementable, facilitated by existing structural components.[116] Yet the museum also performed some activities which even today might be seen as unusual. These activities by contrast were less easily implementable with its existing structural components, requiring scaffolding discussed above to facilitate them.[117]

A third non-public-facing civic service undertaken by the Imperial War Museum was its programme of voluntarily returning various collection items for use in the war. By June 1940, the British Expeditionary Force had been evacuated from continental Europe following the German military's rapid advance into France and the Low Countries. It left behind a considerable amount of equipment, compromising the army's ability to defend the British Isles. The government therefore instigated an emergency re-equipment programme which included accepting *matériel* gifted by the Imperial War Museum: unwanted duplicate items and items considered culturally insignificant.[118] This is a good example of required capability being extended with the aid of scaffolding.[119] The civic service which the Imperial War Museum now found itself performing had evolved beyond the museum's established organisational model. In this example, rather than preserving certain collection items, it returned them to front-line service. Nothing like this had been undertaken by the museum before the Second World War. As such, it required the formation of provisional structuration measures to bridge the philosophical-experiential void and facilitate the necessary outcome.

In August 1940, the Ministry of Supply accepted from the Imperial War Museum ten observation telescopes, two theodolites, one alidade and two dial sights.[120] The British Army also accepted forty steel helmets and four artillery pieces.[121] Later, in October 1940, the Ministry of Supply accepted a second alidade.[122] Other examples following the invasion scare included 14 artillery pieces accepted by the Royal Navy during November 1940–February 1941, an unspecified number of German dial sights by the Ministry of Supply for recycling in November 1941, and surgical supplies and crutches by the nearby St Thomas' Hospital, London, in June 1947.[123] The unpublished wartime history also indicates that the Royal Veterinary College, also London, accepted veterinary equipment in July 1940.[124]

No practice- or scholarly based discussion on this 'industrial reclamation', as the activity is termed here, has been discovered in prevailing museum literature. That could be because such ideas challenge long held assumptions about the irreversibility of the symbolic transition by a collection item from functional object to museum object.[125] Objects in collections are routinely exchanged between museums. They are also disposed of when deterioration prevents them from properly representing the thing they once comprised. But rarely are objects deaccessioned and restored to their pre-accession utility. This act conceivably undermines a museum's rationale of preserving material culture. Granted, certain ethnographic cultural religious items are sometimes restituted for use again amongst their indigenous communities. In such instances however, the objects have never actually been stripped of their function by the peoples which originally used and possessed them.[126] Through breaking with this convention, the Imperial War Museum further evidences the abstract museum and museum sector's reflexiveness to the lived environment. Even the supposedly most sacrosanct institutional values can be temporarily disregarded or adjusted when external pressure mounts up against them, caused by societal value shifts. In doing so, it further evidences the resilience of the museum concept.[127] Some museum advocates may argue that a museum which carries out industrial reclamation as here effectively undermines their primary purpose for existence. Such an argument would have merit if it was applied in a context of peace and the normative societal assumptions that predominate therein. Over 1939–1945 however, the Imperial War Museum was not in a context of peace. Rather, as discussed above, it was in a context of total war. This necessitated an extraordinary response from the museum, indeed all individuals, communities and organisations, which, through regulative, normative and cultural-cognitive institutionalisation, frequently transgressed normative assumptions regarding social, political, economic and cultural life.

Allowing Use of Its Building

The fourth and perhaps simplest way that the Imperial War Museum delivered on a non-public-facing civic service was by sharing its building with other organisations. This is another good example of required capability needing the support of scaffolding.[128] The most visually striking and locally relevant sharing arrangement occurred through it hosting a barrage balloon and its crew of Royal Air Force personnel. Barrage balloons were ground-tethered, lighter-than-air, gas-filled bags which prevented dive bombing and strafing by enemy aeroplanes.[129] The museum provided billets for around 20 crewmembers as early as the year 1938 until 1944.[130] Another such way that the Imperial War Museum supported the local community was by storing emergency feeding equipment for the Ministry of Food.[131] Initially kept in the Naval Gallery, it eventually spread over multiple rooms and floors across the premises. The museum's staff were constantly needed to help maintain the store. As the Imperial War Museum became increasingly damaged during the war however, the equipment became more exposed. Therefore, by September 1942, the store was relocated.[132]

Regionally and nationally relevant war work took place at or operated from the Imperial War Museum as well. Of all the organisations utilising this space, the Ministry of Works became the biggest user and beneficiary. For example, throughout September 1941-August 1943, it maintained a busy vehicle and firefighting equipment repair station.[133] Hence, the premises became quite congested over this period, with increasingly larger pieces of equipment being sent. Eventually, even huge green American Red Cross

Clubmobile buses, renowned for their in-built doughnut making machines, started arriving.[134] This development finally convinced the Ministry of Works that the museum was not the right place to undertake such repairs. Accordingly, the vehicle section eventually transferred elsewhere.[135] Other war work performed by the Ministry of Works included scientific research into gas filtration.[136] Poison gas never featured on the battlefields of the Second World War as it had during the First World War. But the government was not complacent. On the contrary, defensive preparations against gas attacks received considerable attention.[137] Consequently, the Ministry of Works established a laboratory to research methods of gas filtration. This was originally established in the museum's basement during September 1941. By June 1942 it had moved to and expanded on the west wing of B floor during May 1942.[138] The facility then remained there until December 1944 before finally being dismantled in February 1946.[139]

Conclusion

The Second World War initially caused the British government to order the closure of non-essential gathering places, including museums. This disrupted the Imperial War Museum considerably, which found itself prohibited from operating and serving the public as before in expected, excepted ways. But when the dangers posed by war did not initially materialise, the government's order was relaxed, allowing museums and other cultural venues to reopen, albeit with certain constraints. The Imperial War Museum took full advantage of this U-turn. Despite limited resources and restrictions on admission, it responded by curating a fresh, makeshift public exhibition that interpreted the new national situation to an anxious, fretful public, deploying improvised, creative workarounds in the process. This successfully restored some pre-war civic service provision. The eventual onset of regular bombing in September 1940, however, forced the Imperial War Museum to close again until the European conflict had concluded, deepening the disruption it encountered. Thus, the museum was forced to largely withdraw from public gaze during the war years that followed. Yet, as the above examples of other public-facing and non-public-facing activities performed at the time show, it did not withdraw from providing a civic service. Rather, the museum continued maintaining a civic service via other means by drawing on existing strengths. Indeed, it museum became a focal point for broader civic service activities. While the collection was researched or loaned out towards local and national initiatives connected with strengthening and sustaining the country's strategic position, its building played host to equally important local and national initiatives connected with maintaining people's security and welfare.

The current chapter instantiates the idea that organisational crisis management and resilience are two different if not mutually exclusive concepts introduced in Chap. 2. It demonstrates the crucial link between a museum's ability to develop and draw on resilient attributes and to resolve crises. Those attributes exemplified also illustrate the three enabling dimensions of museum resilience discussed above: structures, people and relationships. These included infrastructure and assets, human resources and organisational culture, amongst others. Moreover, it shows that although no attribute suffices resilience alone, some attributes have greater significance than others, depending on the situation being confronted.

The Imperial War Museum overcame its crisis of inactivity during the Second World War by relying heavily on three contributory sources of resilience which together spanned each aforementioned enabling dimension. The first source came from the museum's building and collection. These adaptable resources equipped it with the means to continue delivering a civic service in the face of restrictions by accommodating revised or wholly new activities and practices. The second source came from the museum's staff. Its workforce was empowered to utilise their skills, knowledge and experience of the museum and its work towards augmenting the museum's output *pro re nata*, including performing bricolage and deploying scaffolding. Working the building and collection, they successfully devised, assumed or supported activities which satisfied prevailing societal requirements, even if these diverged from normal operations as evidenced by the work outlined above. The public-facing examples reflected a public need for cultural nourishment and information; the non-public-facing examples reflected a public expectation of engagement in the war effort. And the third source was the museum's network: its place in the national government ecology, relationships with local government, and ability to forge connections within and outwith government. By being a nationally funded institution with links across the establishment, the museum frequently received opportunities and instructions towards performing or otherwise facilitating and supporting civic service activities for the duration of the emergency.

With links drawn between organisational crisis management and resilience, the next chapter examines a second crisis-conducive situation which the Imperial War Museum encountered during the Second World War era.

This was the situation brought on by the various aerial attacks against London from September 1940 until and including May 1945, requiring a particular defensive strategy in being overcome. It would be one which tested the museum, its building, collections and staff, at the limits of their durability and endurance, requiring widespread recourse to resilience exemplified here.

Notes

1. Neil Stammers, *Civil Liberties in Britain During the 2nd World War: A Political Study* (Abingdon: Routledge, 2023).
2. See Richard Overy, *The Bombing War: Europe, 1939–1945*, pbk edn (London: Penguin, 2014); Ina Zweiniger-Bargielowska, *Austerity in Britain: Rationing, Controls, and Consumption, 1939–1955* (Oxford: Oxford University Press, 2000).
3. Catherine Pearson, *Museums in the Second World War: Curators, Culture and Change*, ed Suzanne Keene (Abingdon: Routledge, 2017); Debbie Whittaker, 'Museums at War: The Impact of the Second World War on the National Museums of London, with Particular Reference to the Imperial War Museum, National Gallery, Victoria and Albert Museum and the British Museum' (master's dissertation, University of Leicester, 2010).
4. Christine M. Pearson and Ian I. Mitroff, 'From Crisis Prone to Crisis Prepared: A Framework for Crisis Management', *The Executive* 7 (1993): 52–53.
5. Michael Wayne Sherraden and Donald J. Eberly, 'Why National Service?', in *National Service: Social, Economic, and Military Impacts*, ed Michael Wayne Sherraden and Donald J. Eberly (Elmsford, NY: Pergamon, 1982), 3.
6. James L. Perry and Ann Marie Thomson, *Civic Service: What Difference Does It Make?* (Abingdon: Routledge, 2015), xiv–xv.
7. For example, David C. Leege, 'Catholics and the Civic Order: Parish Participation, Politics, and Civic Participation', *The Review of Politics* 50 (1988): 704–736; Catherine M. Fairley, 'Volunteerism, Youth and the Transition to Democracy in Malawi', *VOSESA Focus* 1 (2006): http://www.vosesa.org.za/focus/vol1_no3/index.html?article_3.html~content (18 March 2024); Rebecca Nesbit and David A. Reingold, 'Soldiers to Citizens: The Link between Military Service and Volunteering', *Public Administration Review* 71 (2011): 67–76.
8. Richard Danzig and Peter Szanton, *National Service: What Would It Mean?* (Lexington, MA: Lexington Books, 1986), 10.

9. Peter Latchford, *The Future of Civic Museums: A Think Piece* (n.place: Black Radley, 2018), 5, https://www.nationalmuseums.org.uk/media/documents/publications/civic_museums_think_piece.pdf (18 March 2024).
10. Simon Hornblower, Antony Spawforth and Esther Eidinow, eds, *The Oxford Classical Dictionary*, 4th edn (Oxford: Oxford University Press, 2012), s.v. 'civitas'.
11. Chris Brink, *The Soul of a University: Why Excellence Is Not Enough* (Bristol: Bristol University Press, 2018), 326.
12. For example, Robert R. Janes, *Museums in a Troubled World: Renewal, Irrelevance or Collapse?* (Abingdon: Routledge, 2009); Stephanie Lessans Geller and Lester M. Salamon (with the technical assistance of Kasey L. Mengel), 'Museums and Other Nonprofits in the Current Recession: A Story of Resilience, Innovation, and Survival', *Journal of Museum Education* 35 (2010): 129–140; Robert R. Janes, 'Museums and the End of Materialism', in *The Routledge Companion to Museum Ethics: Redefining Ethics for the Twenty-First Century Museum*, ed Janet Marstine (Abingdon: Routledge, 2011), 54–69; Robert R. Janes, *Museums Without Borders: Selected Writings of Robert R. Janes* (Abingdon: Routledge, 2016); Joshua Decter, 'Will There Still Be a Future When the Museum of the Future Arrives?', in *The Future of Museums*, ed Gerald Bast, Elias G. Carayannis and David F. J. Campbell (Cham: Springer International, 2018), 15–36; Mara Cerquetti and Eleonora Cutrini, 'Structure, People and Relationships: A Multidimensional Method to Assess Museum Resilience', *Nonprofit and Voluntary Sector Quarterly* 52 (2023): 130–152.
13. James J. Leflar and Marc H. Siegel, *Organizational Resilience: Managing the Risks of Disruptive Events – A Practitioner's Guide* (Boca Raton, FL: CRC, 2013), 11; Ran S. Bhamra, Kevin Burnard and Samir S. Dani, 'Resilience: A Concept, a Literature Review and Future Directions', in *Organisational Resilience: Concepts, Integration and Practice*, ed Ran Bhamra (Boca Raton, FL: CRC, 2015), 21; McCarthy, Mark Collard and Michael Johnson, 'Adaptive Organizational Resilience: An Evolutionary Perspective', *Current Opinion in Environmental Sustainability* 28 (2017): 33–40.
14. Erika Anderson, 'Learning to Learn', *Harvard Business Review* 94 (2016): 99.
15. Keith Shaw and Louise Maythorne, 'Managing for Local Resilience: Towards a Strategic Approach', *Public Policy and Administration* 28 (2011): 52; see Brigit Maguire and Sophie Cartwright, *Assessing a Community's Capacity to Manage Change: A Resilience Approach to Social Assessment* (Canberra, ACT: Bureau of Rural Science, 2008), https://

www.agriculture.gov.au/sites/default/files/abares/documents/dewha-resilience-sa-report-final-4.pdf (18 March 2024).

16. Ron Martin and Peter Sunley, 'On the Notion of Regional Economic Resilience: Conceptualization and Explanation', *Journal of Economic Geography* 15 (2015): 4.
17. Shaw and Maythorne, 'Managing, 46.
18. Ibid., 52.
19. Martin and Sunley, 'Regional Economic Resilience', 4.
20. Shaw and Maythorne, 'Managing', 46.
21. Mike Raco and Emma Street, 'Resilience Planning, Economic Change and The Politics of Post-recession Development in London and Hong Kong', *Urban Studies* 49 (2011): 1069.
22. Richard J. T. Klein, Robert J. Nicholls and Frank Thomalla, 'Resilience to Natural Hazards: How Useful Is This Concept?', *Environmental Hazards* 5 (2003): 42.
23. Cerquetti and Cutrini, 'Structure', 135.
24. Charlotte Brown, Erica Seville and John Vargo, 'Measuring the Organizational Resilience of Critical Infrastructure Providers: A New Zealand Case Study', *International Journal of Critical Infrastructure Protection* 18 (2017): 38.
25. Thomas W. Brit and Gargi Sawhney, 'Resilience Capacity, Processes and Demonstration at the Employee, Team and Organizational Levels: A Multilevel Perspective', in *Research Handbook on Organizational Resilience*, ed Edward H. Powley, Brianna Barker Caza and Arran Caza (Cheltenham: Edward Elgar, 2020), 15.
26. Sonia McManus, Erica Seville, John Vargo and David Brunsdon, 'Facilitated Process for Improving Organizational Resilience', *Natural Hazards Review* 9 (2008): 81–90.
27. Larry A. Mallak, 'Measuring Resilience in Health Care Provider Organizations', *Health Manpower Management* 24 (1998): 148–152; Cynthia A. Lengnick-Hall and Tammy E. Beck, 'Resilience Capacity and Strategic Agility: Prerequisites for Thriving in a Dynamic Environment', in *Resilience and Engineering*, Resilience Engineering Perspectives 2, ed Christpher P. Nemeth, Erik Hollnagel and Sidney Dekker (Farnham: Ashgate, 2009), 39–70.
28. John F. Horne III and John E. Orr, 'Assessing Behaviours That Create Resilient Organizations', *Employment Relations Today* 24 (1998): 39; Cynthia A. Lengnick-Hall, Tammy E. Beck and Mark L. Lengnick-Hall, 'Developing A Capacity for Organizational Resilience Through Strategic Human Resource Management', *Human Resource Management Review* 21 (2011): 243–255.

29. Anna Tulliach, 'Best Practice in Museum Partnerships: Bologna and its Museum Networking', *Museological Review* 21 (2017): 8–22.
30. Graham Matthews, Yvonne Smith and Gemma Knowles, *Disaster Management in Archives, Libraries and Museums* (Aldershot: Ashgate, 2009), 176–177; Mara Cerquetti, 'Vingt ans après. Do Museum Networks Really Work? A Focus on the Inland Areas of the Marche Region's "Seismic Crater" (Italy)' *Il Capitale Culturale. Studies on the Value of Cultural Heritage* 19 (2019): 27–51.
31. Janes, *Troubled World*, 121–146.
32. Ibid., 142; see Stephen Weil, *Making Museums Matter* (Washington, DC: Smithsonian Institution Press, 2002).
33. See Sheila Watson, 'Museums and their Communities', in *Museums and their Communities*, ed Sheila Watson (Abingdon: Routledge, 2007), 1–23.
34. Cerquetti and Cutrini, 'Structure', 135.
35. Samuel J. Redman, *The Museum: A Short History of Crisis and Resilience* (New York, NY: New York University Press, 2022).
36. S. F. Markham, *A Report on the Museums and Art Galleries of the British Isles (Other Than the National Museums)* (Edinburgh: Carnegie United Kingdom Trustees, 1938), 174.
37. Jörn Weingärtner, *The Arts as a Weapon of War: Britain and the Shaping of National Morale in World War II*, rev. pbk edn (London: I. B. Tauris, 2012), 50; see Andrew Davies, 'The War Years', in *Theatre and War, 1933-1945: Performance in Extremis*, ed Michael Balfour (New York, NY: Berghahn, 2001), 54–55; Anthony Aldgate and Jeffrey Richards, *Britain Can Take It: The British Cinema in the Second World War*, new edn (London: I. B. Tauris, 2007), 1; Pearson, *Museums*, 67.
38. Nicholas Atkin, Michael Biddiss and Frank Tallett, *The Wiley-Blackwell Dictionary of Modern European History Since 1789* (Chichester: John Wiley, 2011), s.v. 'phoney war'; Robert Mackay, *Half the Battle: Civilian Morale in Britain during the Second World War* (Manchester: Manchester University Press, 2002), 50.
39. Weingärtner, *Arts*, 50–51; The National Archives (TNA), EB 3/17, Markham to the Financial Secretary to the Treasury, 2 November 1939; TNA, EB 3/17, 'The question of the reopening of the National Museums and Galleries', 20 December 1939.
40. 'The Museums Reopen', *Museums Journal* 39 (1940): 454–455.
41. IWM, Museum Archive (MA), EN1/1/COB/049/4, Board of Trustees, 7 December 1939, 3; IWM, MA, EN2/1/COB/001/1, Standing Committee, 11 January 1940, 1.
42. Jamie Watters, *The Business Continuity Management Desk Reference: Guide to Business Continuity Planning, Crisis Management and IT Disaster Recovery* (n.place: Business Leverage, 2010), 9.

43. Joseph Mathenge, 'Business Continuity vs Business Resiliency: What's the Difference?', The Business of IT Blog, BMC, https://www.bmc.com/blogs/business-continuity-vs-resiliency/ (18 March 2024); see Rachelle Loyear, *The Manager's Guide to Simple, Strategic, Service-Oriented Business Continuity*, ed Kristen Noakes-Fry (n.place: Rothstein, 2017), chapt. 8, https://www.rothstein.com/product/simple-strategic-service-oriented-business-continuity-management/ (18 March 2024).
44. Ernest Blaikley, 'Reopening of the Imperial War Museum', *Museums Journal* 40 (1940): 8.
45. Ibid.
46. Ibid.
47. IWM, MA, EN1/1/COB/049/4, Board of Trustees, 7 December 1939, 3.
48. Blaikley, 'Reopening', 8; see Gordon Landsborough, *The Battle of the River Plate: The First Naval Battle of the Second World War* (Barnsley: Frontline, 2016).
49. Blaikley, 'Reopening', 8; see David Faber, *Munich: The 1938 Appeasement Crisis* (London: Pocket, 2009).
50. '"Peace in Our Time" in the War Museum', *Daily Mail*, 7 February 1940; 'Framed', *Daily Mirror*, 7 February 1940.
51. Mollie Panter-Downes, *London War Notes, 1939–1945*, ed William Shawn (London: Longman, 1972), 45.
52. Blaikley, 'Reopening', 8–9; 'The Imperial War Museum', *Times*, 26 January 1940.
53. Blaikley, 'Reopening', 8.
54. Ibid.
55. Ibid., 9.
56. Martin Gilbert, *Winston S. Churchill, 1922–1939: Prophet of Truth* (London: William Heinemann, 1976).
57. 'The Imperial War Museum Reopens', *Illustrated London News*, 3 February 1940; see 'Mr. Churchill in Manchester', *Manchester Guardian*, 29 January 1940.
58. James M. Kendra and Tricia Wachtendorf, 'Elements of Resilience After the World Trade Center Disaster: Reconstituting New York City's Emergency Operations Centre', *Disasters* 27 (2003): 42.
59. Claude Lévi-Strauss, *The Savage Mind* (Chicago, IL: University of Chicago Press, 1966).
60. Stephanie Duchek, 'Organizational Resilience: A Capability-based Conceptualization', *Business Research* 13 (2020): 228–229; Gene I. Rochlin, 'Informal Organisational Networking as a Crisis Avoidance Strategy: US Naval Flight Operations as a Case Study', *Industrial Crisis Quarterly* 3 (1989): 159–176; Karl E. Weick, 'The Collapse of

Sensemaking in Organizations: The Mann Gulch Disaster', *Administrative Science Quarterly* 38 (1993): 628–652; Mathilde Bourrier, 'Organizing Maintenance Work at Two Nuclear Power Plants', *Journal of Contingencies and Crisis Management* 4 (1996): 104–112; Karl E. Weick, Kathhleen M. Sutcliffe and David Obstfeld, 'Organizing for High Reliability: Processes of Collective Mindfulness', in *Research in Organizational Behaviour*, ed Barry M. Staw and Robert I. Sutton, 21 (Greenwich, CT: JAI, 1999), 81–123.

61. Blaikley, 'Reopening', 8.
62. 'Imperial War Museum', *Times*.
63. Emlyn H. Koster and Stephen H. Baumann, 'Liberty Science Center in the United States: A Mission Focussed on External Relevance', in *Looking Reality in the Eye: Museums and Social Responsibility*, ed Robert R. Janes and Gerald T. Conty (Calgary, AB: University of Calgary Press, 2005), 86.
64. Weil, *Making Museums Matter*, 70; Rhiannon Mason, 'Museums, Galleries and Heritage: Sites of Meaning-making and Communication', in *Heritage, Museums and Galleries: An Introductory Reader*, ed Gerard Corsane (Abingdon: Routledge, 2005), 200–214.
65. David Farrell-Banks, *Affect and Belonging in Political Uses of the Past* (Abingdon: Routledge, 2023), 3–4.
66. David Welch, 'Propaganda, Definitions of', in *Propaganda and Mass Persuasion: A Historical Encycloedia, 1500 to the Present*, ed Nicholas J. Cull, David Culbert and David Welch (Santa-Barbra, CA: ABC-CLIO, 2003), 318.
67. IWM, MA, EN1/1/REP/032, 'War History of the Imperial War Museum, 1933–1943', 30.
68. IWM, MA, EN1/1/MUS/024/10, Bradley to Rath, 6 September 1933, 1.
69. Panter-Downes, *London War Notes*, 46.
70. Julie Gottlieb, 'Umbrellas, Hats, and Gas Masks: The Objects of the War of Nerves, Britain 1938–1939' (keynote address, Objects In and After Hostility: The Materiality of Conflict, Newcastle upon Tyne, 30–31 March 2017).
71. Amy Bell, 'Landscapes of Fear: Wartime London, 1939–1945', *Journal of British Studies* 48 (2009): 153–75.
72. Paul J. Silvia, 'Curiosity and Motivation', in *The Oxford Handbook of Human Motivation*, ed Richard M. Ryan (Oxford: Oxford University Press, 2012), 157–166.
73. George Loewenstein, 'The Psychology of Curiosity: A Review and Reinterpretation', *Psychological Bulletin* 116 (1994): 87.
74. David H. Barlow, *Anxiety and Its Disorders: The Nature and Treatment of Anxiety and Panic*, 2nd edn (New York, NY: Guilford, 2004), 256.

75. Shaw and Maythorne, 'Managing', 43–65.
76. Diana Condell, 'The History and Role of the Imperial War Museum', in *War and the Cultural Construction of Identities in Britain*, ed Barbara Korte and Ralf Schneider (Amsterdam: Rodopi, 2002), 31.
77. Hannah Paddon, *Redisplaying Museum Collections: Contemporary Display and Interpretation in British Museums* (Farnham: Ashgate, 2014).
78. Simon J. Knell, Suzanne Macleod and Sheila Watson, 'Introduction', in *Museum Revolutions: How Museums Change and Are Changed*, ed Simon J. Knell, Suzanne Macleod and Sheila Watson (Abingdon: Routledge, 2007), xix–xxvi; Gail Anderson, 'A Framework: Reinventing the Museum', in *Reinventing the Museum: The Evolving Conversation on the Paradigm Shift*, ed Gail Anderson (Lanham, MD: AltaMira, 2012), 1–9.
79. Janes, *Troubled World*, 141.
80. IWM, MA, EN1/1/REP/032, 'War History [...], 1933–1943', 10.
81. Ibid.
82. IWM, MA, EN2/1/MUS/003/1b, Griffiths to Bradley, 11 December 1939; IWM, MA, EN2/1/MUS/003/1b, Bradley to Griffiths, 6 March 1940; IWM, MA, EN2/1/MUS/003/1b, Griffiths to Bradley, 19 March 1940.
83. IWM, MA, EN2/1/MUS/003/1b, Bradley to Griffiths, 26 March 1940.
84. IWM, MA, EN1/1/REP/032, 'War History [...], 1933–1943', 10.
85. Gareth R. Jones, *Organizational Theory: Text and Cases*, 3rd ill. edn (Upper Saddle River, NJ: Pretence-Hall, 2001), 2.
86. Peter Berger and Thomas Luckmann, *The Construction of Reality: A Treatise in the Sociology of Knowledge* (London: Penguin, 1991), 93; Tim Hallett and Marc J. Ventresca, 'Inhabited Institutions: Social Interactions and Organizational Forms in Gouldner's "Patterns of Industrial Bureaucracy"', *Theory and Society* 35 (2006): 213–236.
87. Timothy L. J. Ferris, Scott Jackson; Keith D. Willett; Victoria A. Hailey and Eric Specking, 'System Resilience: Application Domain Perspectives', *INCOSE International Symposium* 29 (2019): 1094.
88. Charles Vincent, Susan Burnett and Jane Carthey, *The Measurement and Monitoring of Safety: Drawing Together Academic Evidence and Practical Experience to Produce a Framework for Safety Measurement and Monitoring* (London: Health Foundation, 2013), 56.
89. IWM, MA, EN2/1/ENQ/001, see documents therein.
90. IWM, MA, EN2/1/ENQ/001, Males to IWM, 28 March 1940.
91. IWM, MA, EN2/1/ENQ/024, Richardson to IWM, 15 June 1943.
92. IWM, MA, EN2/1/ENQ/001, Garrett to IWM, 17 October 1944, 1–2.
93. Ibid., 2.

94. IWM, MA, EN2/1/ENQ/001, Bradley to Males, 2 April 1940; IWM, MA, EN2/1/ENQ/024, Bradley to Richardson, 24 June 1943; IWM, MA, EN2/1/ENQ/001, Bradley to Garrett, 19 October 1944.
95. Wanda J. Orlikowski, 'Material Knowing: The Scaffolding of Human Knowledgeability', *European Journal of Information Systems* 15 (2006): 461–462.
96. Andy Clark, 'Magic Words: How Language Augments Human Computation', in *Language and Thought: Interdisciplinary Themes*, ed Peter Carruthers and Jill Boucher (Cambridge: Cambridge University Press, 1998), 163.
97. John Child, *Organization: Contemporary Principles and Practice* (Malden, MA: Blackwell, 2005), 6–9.
98. IWM, MA, EN1/1/REP/032, 'War History [...], 1933–1943', 27.
99. IWM, MA, EN2/1/ENQ/024, Bradley to Buckingham Palace, 22 June 1943.
100. IWM, MA, EN2/1/ENQ/024, Miéville to Bradley, 23 June 1943; IWM, MA, EN2/1/ENQ/024, Bradley to Miéville, 24 June 1943; IWM, MA, EN2/1/ENQ/024, Bradley to Richardson, 24 June 1943; IWM, MA, EN2/1/ENQ/024, Bradley to Richardson, 19 July 1943.
101. Watters, *Continuity Management*, 9.
102. Ferris et al., 'System Resilience', 1094.
103. Franz Uhle-Wettler, 'War', in *Brassey's Encyclopedia of Military History and Biography*, ed Franklin D. Margiotta (Washington, DC: Brassey's, 1994), 1047.
104. Standing Commission on Museums and Galleries, *Third Report: The War Years and After* (London: HMSO, 1948).
105. Janes, *Troubled World*, 141–145.
106. Sharon Heal, 'Editorial', *Museums Journal* 114, no. 4 (2014): 4.
107. Jones, *Organizational Theory*, 2.
108. IWM, MA, EN1/1/REP/032, 'War History [...], 1933–1943', 26–29.
109. Kendra and Wachtendorf, 'Elements of Resilience', 42.
110. IWM, MA, EN1/1/REP/032, 'Imperial War Museum: 21st Annual Report of the Director-General to the Board of Trustees', 1; Committee of Public Accounts, *Special Report and First, Second and Third Reports from the Committee of Public Accounts Together with the Proceedings of the Committee, Minutes of Evidence, Appendices and Index* (London: HMSO, 1957), 17; Joyce Apsel, *Introducing Peace Museums* (Abingdon: Routledge, 2016), 12.
111. Ritchie Calder, 'Bernal at War', in *J. D. Bernal: A Life in Science and* Politics, ed Brenda Swann and Francis Aprahamian (London: Verso, 1999), 165–166.

112. TNA, HO 196/4, 'Air Raid Damage in London, 1914–1918: Records and Photographs in the Imperial War Museum', 27 November 1939.
113. Ibid., 2; IWM, MA, EN1/1/REP/032, 'War History [...], 1933–1943', 26–27.
114. IWM, MA, EN1/1/REP/032, 'War History [...], 1933–1943', 26–29; Norman Longmate, *How We Lived Then: A History of Every Day Life During the Second World War* (London: Pimlico, 2002), 382; Rosalind Watkiss Singleton, "Doing Your Bit': Women and the National Savings Movement in the Second World War', in *The Home Front in Britain: Images, Myths and Forgotten Experiences since 1914*, ed Maggie Andrews and Janis Lomas (Basingstoke: Palgrave Macmillan, 2014), 217–231.
115. Kendra and Wachtendorf, 'Elements of Resilience', 42.
116. Child, *Organization*, 9–11.
117. Orlikowski, 'Material Knowing', 461–462.
118. IWM, MA, EN1/1/REP/032, 'War History [...], 1933–1943', 7; Julian Thompson, *Dunkirk: Retreat to Victory* (London: Pan, 2009), 297; Terry Charman, '"A Museum of Man's Greatest Lunatic Folly": The Imperial War Museum and its Commemoration of the Great War, 1917–2008', in *A Part of History: Aspects of the British Experience of the Western Front* (London: Continuum, 2008), 103.
119. Ferris et al., 'System Resilience', 1094; Orlikowski, 'Material Knowing', 461–462.
120. IWM, MA, EN2/1/DIS/001, unidentifiable individual [Assistant Director of Instrument Production] to Bradley, 15 August 1940, plus overleaf receipt signed by Johnson, 16 August 1940.
121. IWM, MA, EN2/1/DIS/001, Naley to Bradley, countersigned by Derham, 14 August 1940; IWM, MA, EN2/1/DIS/001, Hammond to Bradley, 26 July 1940; IWM, MA, EN2/1/DIS/001, list of weapons resituated with the following hand-written note underneath: 'Taken away 16. 8. 40. See letter from War Office dated 26 July 1940. Ref. 57/M.A./4745.'
122. IWM, MA, EN2/1/DIS/001, Manualt to Bradley, 30 October 1940.
123. IWM, MA, EN2/1/GOV/011, see documents therein; IWM, MA, EN2/1/DIS/001, unidentifiable individual [Assistant Director of Instrument Production] to Bradley, 11 November 1941; IWM, MA, EN2/1/DIS/001, Pearson to Bradley, 17 June 1947.
124. IWM, MA, EN1/1/REP/032, 'War History [...], 1933–1943', 7.
125. Hilde S. Hein, *The Museum in Transition: A Philosophical Perspective* (Washington, DC: Smithsonian Institution Press, 2000), 54–57.
126. Piotr Bienkowski, 'Authority and the Power of Place: Exploring the Legitimacy of Authorized Voices in the Restitution Discourse', in

Museums and Restitution, ed Louise Tythacott and Kostas Arvanitis. Farnham: Ashgate, 2014, 37–52.

127. Decter, 'Will There Still', 17.
128. Ferris et al., 'System Resilience', 1094; Orlikowski, 'Material Knowing', 461–462.
129. Franklin J. Hillson, 'Barrage Balloons for Low Level Air Defence', *Airpower Journal* 3 (1989): 31–32.
130. IWM, MA, EN1/1/REP/032, 'War History [...], 1933–1943', 33.
131. Ibid.
132. Ibid.
133. Ibid., 34.
134. James H. Madison, *Slinging Doughnuts for the Boys: An American Woman in World War II* (Bloomington, IN: Indiana University Press, 2007), 25.
135. IWM, MA, EN1/1/REP/032, 'War History [...], 1933–1943', 34.
136. Ibid.
137. Brown, Frederic J., *Chemical Warfare: A Study in Restraint*, edn with new intro (New Brunswick, NJ: Transaction, 2006).
138. IWM, MA, EN1/1/REP/032, 'War History [...], 1933–1943', 34.
139. IWM, MA, EN1/1/REP/032, 'War History of the Imperial War Museum: Part II, January 1944 to the Reopening of the Museum on 27th November 1946', 13.

CHAPTER 5

Safeguarding the Collection, 1933–1950

Introduction

So far, this study has provided a detailed if foundational explanation of what crises are and how they can be resolved, focussing on the organisational and particularly museum context. The discussion hitherto has emphasised their propensity to cause and signify disruption in crisis objects. It has also emphasised the organisational imperative of fostering and drawing on resilience when resolving crises. Now, the study compares and contrasts two different crisis typologies. The current chapter uses the crisis of physical destruction at the Imperial War Museum during the Second World War to develop a typology of the concept caused by tangible pressures which museums should, in those instances, respond defensively against. This corresponds with the concept's classical reading, introduced in Chap. 2, where an object's continuance is poised precariously between positive (sustainable) and negative (unsustainable) states. The chapter shows the susceptibility of museums to such crises. It also discusses the methodology which museums should follow when resolving them.

The historical context of the current chapter is the campaign of aerial attacks ordered by the German High Command against London during the Second World War. These attacks materialised variously, initially as piloted bombing raids, and later as unpiloted V-1 flying bomb and V-2 rocket strikes.[1] The Imperial War Museum suffered greatly from damage caused during each phase of the campaign. Indeed, in 1941, Ernest

P. W. Deans, *Crisis, Reinvention and Resilience in Museums*,
https://doi.org/10.1007/978-3-031-67806-6_5

Blaikley, the museum's art curator, described utterly torrid working conditions there: privation, danger from air raid alerts, and exposure to the elements through building degradation.[2] The first post-war report by the Standing Commission on Museums and Galleries commented on the 'exceptional' nature of the toll which the museum paid compared with similar national institutions.[3]

The current chapter draws on two crisis concepts to explicate the crisis of physical destruction brought about by the aerial attacks against London. One is the 'surface-defensive crisis', the first novel crisis typology conceived and theorised towards this research, established below, which characterises the crisis-conducive situation confronting the museum. The other is Christine M. Pearson and Ian I. Mitroff's five-stage framework for crisis management, previously introduced in Chap. 2.[4] The latter establishes five different stages comprising the crisis management cycle, commencing with the search for crisis signals and concluding with what can be usefully learnt from the process. It has been deployed here to structure the narrative and analysis of the aerial attacks and their impact on the museum.

The current chapter addresses the above over three substantive sections. The first section introduces and explicates surface-defensive crisis. It sets out the typology's premise and signification. The second considers the response by the Imperial War Museum to the crisis of physical destruction. It explores the museum's methodology for resolving this situation, including detecting and preparing for, mitigating, and recovering and learning from the disruption. And the third discusses how the aerial attacks and the counter measures deployed by the museum comprise an example of surface-defensive crisis and crisis management. It synthesises ideas from the concept with findings from the case study.

Surface-Defensive Crisis

The surface-defensive crisis is an organisational crisis typology predicated on two descriptors. The first descriptor, 'surface', establishes the practical and critical significance of all applicable crisis situations as well as the implications which these crises have for an organisation. It highlights the material, that is the physical, dimension which they will manifest in and cause impact to organisations. It also conveys their point of focus as an analytical tool, these being on practical everyday organisational operations. The second descriptor, 'defensive', establishes the necessary organisational response in dealing with all applicable crises. It references the conservative posture of the mitigating measures that should be deployed by organisations against them. It also signifies the various dangers which organisations can face on being impacted. These can include anything physical and tangible, from natural, environmental occurrences to intentional, human acts.[5]

A surface-defensive crisis impacts museums superficially. That is, it can be felt chiefly through damage to infrastructure and disruption to processes and output, meaning wider discourses about organisational stability and sustainability bear little if any relevance. Indeed, surface-defensive crises do not arise out of the dynamics that surround, buffet and change the relationship between a museum and its superordinate social system, such as social, political, cultural or economic shifts denoting broader societal trends. Rather, they arise out of disruption to daily operations, including the assets which facilitate them. Hence, surface-defensive crises comprise practical, rather than theoretical, crises, and are the most likely reason why sufficiently valued and resourced museums experience difficulty and fail.

The aim of surface-defensive crisis management is to maintain the *status quo*. This can be achieved by averting a crisis before it strikes or by mitigating the crisis' effects. As Fig. 5.1 illustrates, surface-defensive crisis management constitutes an act of resisting the 'crisis conditions'. This renders the surface-defensive crisis analogous to 'disaster', which E. L. Quarantelli has called 'relatively sudden occasions when, because of perceived threats, the routines of collective social units are seriously disrupted and when unplanned courses of action have to be undertaken to cope with the crisis'.[6] Consequently, the surface-defensive crisis can be understood as a typology of crisis that draws on the concept's classical reading.[7] It also overlaps with the recovery conceptualisation of organisational resilience discussed in Chap. 4, where the recourse to resilience involves perseverance through turbulence.[8]

Resolving a Crisis of Physical Destruction at the Imperial War Museum

Detecting the Threat

Surface-defensive crisis management begins for museums, as for all organisations, with the discovery of threats, ideally before disruption becomes reality. Signal detection, the first stage in Pearson and Mitroff's five-stage framework, is the period wherein a museum discovers an existential crisis. It involves undertaking assessments to distinguish between crisis signals

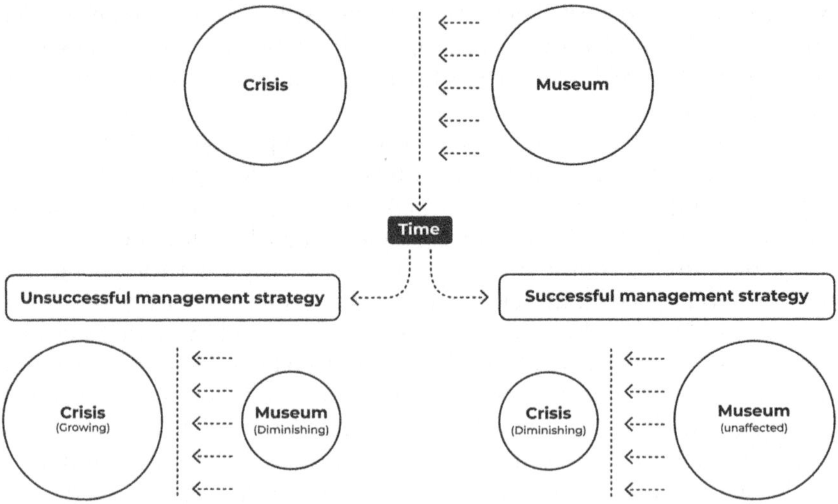

Fig. 5.1 Diagram illustrating two possible outcomes of surface-defensive crisis. Note that the museum which unsuccessfully pushes back against (withstands) environmental turbulence will be diminished, whereas the museum which successfully pushes back will remain fundamentally unaffected

and everyday noise.[9] This task is not easily accomplished however, given the presence of conflicting signals. Indeed, the issue has received attention in economics, where economists have developed tools for analysing crisis potentials to minimise confusion.[10]

In practice, signal detection comprises two tasks. The first task involves analysing all incoming information—which must be systematically collected, processed and disseminated and is essential for coordinating crisis management[11]—to distinguish between red flags and background noise: a process of risk identification and analysis.[12] The term 'red flag' is a metaphor for 'warning sign'. In crisis parlance, red flags could be any phenomenon which signifies something amiss causing unusual data returns.[13] Thus, for museums confronting a surface-defensive crisis, red flags will comprise any indicator of tangible, rather than intangible, problematic phenomena, such as raised humidity levels. The second task involves analysing red flags to try and determine their credibility: evaluating their risk potential.[14]

This is the twofold process wherein crisis contexts are formulated and through which crisis communities arise.[15] As discussed in Chap. 2, crises are conceived by someone and verified by others. The community of the crisis conscious licences itself to delineate the crisis. But not all red flags will be genuine despite contrary perceptions, meaning not all responses will be necessary. Certain red flags may be erroneous, benign or idiopathic.[16] If the wrong judgment call is made, the outcome could cause harm to a museum.[17]

It was signal detection at a governmental level that catalysed preparations by the Imperial War Museum for the surface-defensive crisis of the Second World War. In other words, the British government detected threats, and the museum legitimised them. One red flag was Germany's withdrawal from the League of Nations and Disarmament Conference during November 1933.[18] Another was the advancement in aerial warfare which exacerbated the potential ramifications of the former to the United Kingdom.[19] Aircraft now had significant military capabilities, the Spanish Civil War demonstrating their potential.[20] Compounded by the belief that 'the bomber will always get through',[21] as Prime Minister Stanley Baldwin declared to the House of Commons on 10 November 1932, fears about a sudden, overwhelming 'knockout blow' against London during war preoccupied policy makers over the 1930s as countries rearmed and conflict loomed on the continent.[22]

Through critiquing these red flags, the government sought the Imperial War Museum and other national institutions to prepare for the effects of war over 1933–1934. Specifically, the Office of Works was concerned about the London national museums and their collections should the capital be attacked by air.[23] The First Commissioner of Works therefore convened a conference on the 'Safe Custody of National Art Treasures in the event of War', attended by delegates from all the London national museums. Representing the Imperial War Museum was Leslie Bradley, its then newly promoted Curator and Secretary. At this conference, the delegates discussed their institution's requirements for dealing with the emerging situation.[24]

Preparing for the Onslaught

Once a potential crisis has been detected, the next stage of crisis management, as set out in Pearson and Mitroff's five-stage framework, is to prepare for crisis, which includes applying all available necessary resources and capabilities against the problem. No museum crisis manager should assume that their museum can resolve a crisis without prior preparation. After all, crises are entirely unpredictable. Planning for a range of different crisis scenarios will ensure the museum is in the best possible place to address whatever arises. Effective preparation relies on two tasks being accomplished beforehand.[25] The first task involves identifying the typology of the crisis being confronted.[26] A museum will need to determine, firstly, whether the situation it faces possesses defensive or revolutionary properties and, secondly, can be prevented or not.[27] This is because the nature of the crisis determines the necessary response. The second task involves formulating that response by devising an appropriate strategy to deal with it.[28] A strategy informs all subsequent activities towards preventing potential crisis or weathering the impact where prevention is impossible.[29] Surface-defensive crisis strategies could be preventative, but they could also be mitigative. Whether one or the other ultimately depends on two key factors: the origins of the crisis—what caused it to begin with—and the stage at which the crisis is detected.

The records from the conference on the Safe Custody of National Art Treasures indicate that the delegates perceived a defensive crisis confronting their museums, even if they did not acknowledge that reading of the situation at the time. This is because the recorded conversations articulate a strategy concerned with physically preserving the institutions. If necessary and feasible, the primary action was to evacuate the collections from their respective museums and secure them in safely located refuges.[30]

The Imperial War Museum responded to this situation with preparations constitutive of the agreed strategy. It commenced this work by instigating 'capability-building' measures: generating the knowledge and skills for performing a task.[31] In crisis planning, Stephanie Duchek describes such work as developing 'resources that are necessary in times of crisis (e.g., suitable recovery plans, effective relationships, and mutual understanding)'.[32] Examples discussed below where the institution achieved this include identifying collection items requiring evacuation and securing country houses for their refuge.

Following an interregnum over the winter spanning 1934–1935, which was largely taken up by transferring from South Kensington to Southwark, the Imperial War Museum continued crisis management preparations with focus on 'capacity-building': establishing the resources for performing a task.[33] In particular, the museum enhanced its new building's protective qualities.[34] This is because there was general concern that an orderly evacuation might be impossible.[35] Indeed, the unpublished war history recounts a fear during the earliest preparations that the collection might have needed evacuating 'at very short notice, at great speed—possibly at night—and under fire'.[36] Moreover, that many of its collection items could not be removed.[37] In response, the Office of Works undertook to build a refuge in the museum's basement where collection items could be stored as a last resort.[38]

It is important to note that not all crises are preventable. Some derive from circumstances which develop away from open view. Others, including the example discussed here confronting the Imperial War Museum, derive from circumstances out of organisational reach or beyond organisational control, visible or otherwise.[39] Hence, the benchmark for successful crisis management, including surface-defensive crisis management, is sometimes not predicated on whether a crisis was prevented, but whether the crisis was survived—an equally important outcome.[40] The next subsection considers this latter course and consequence by exploring the museum's work to limit and contain the threat of aerial attack.

Mitigating the Damage

Where a crisis cannot be offset, sooner or later it will strike. When this occurs, crisis management, if conforming to Pearson and Mitroff's five-stage framework, enters the third stage: mitigating the effects. This is where previous crisis planning, discussed above, gets applied against the crisis, with those museums possessing developed and considered plans usually faring better than those which do not.[41] A critique of Pearson and Mitroff's framework points towards mitigation as comprising two tasks.[42] The first task, containment, involves preventing the crisis from spreading, thereby curtailing its damage capability.[43] Historical research by Niall Ferguson on the causes, courses and outcomes of crises shows that the more intricate a system, the greater the risk of collapse through the progressive disruption of connected nodes.[44] Containment thus is particularly important to complex and high value organisations such as museums confronting surface-defensive crisis, where even the slightest spread can cause paralysis and costly damage to assets. The second task, damage limitation, parallels containment, but revolves around limiting the crisis' impact wherever the crisis spreads.[45] Both tasks are undertaken in a context of continual crisis assessment, which involves analysing the crisis with the view to amending the existing crisis management plan as required.[46] Where no crisis management plan exists, crisis assessment should instead take place with the view to determining what resources must be deployed against the threat.[47] Fleet-footedness is essential during crisis mitigation, especially the task of containment. Any delays risk the situation developing beyond control or deteriorating to dangerous levels.[48]

The United Kingdom declared war against Germany on 3 September 1939. But at the Imperial War Museum, measures dealing with the threats posed by the war had already been underway for eleven days. As the international situation intensified, the Home Secretary ordered on 23 August 1939 that the London national museums should initiate their mitigation plans to preclude the possibility of the measures being caught in any opening engagements.[49] Such moves embody 'risk avoidance' strategies and comprise actions aimed at reducing likely and highly impactful disruption.[50] For the Imperial War Museum, this involved gathering together preselected collection items—certain specific artworks and all its photographic albums[51]—by three exits in preparation to be evacuated.[52] Moreover, the artwork was arranged at the exits into three 'classes'. This not only determined their order of evacuation, but also indicates the value they were held in. Those designated as 'Class I' material were scheduled to be evacuated first, followed by those designated as 'Class II', and then those designated as 'Class III'.[53] On 24 August, two government-arranged lorries conveyed material to prearranged country houses. A further three loads went on 25 August.[54]

By 1939, three large country houses owned by Trustees of the Imperial War Museum had been selected as refuges for evacuated collection items. These were Penn House near Amersham in Buckinghamshire, owned by Francis Curzon, 5th Earl Howe; Ramster Hall near Chiddingfold in Surrey, owned by Florence Priscilla Norman, Lady Norman; and Colworth House near Sharnbrook in Bedfordshire, owned by Henry Ludwig Mond, 2nd Baron Melchett.[55] Such properties had positive and negative attributes to them as refuges for evacuated collections. On one hand, country houses were located outside key target areas and often diffusely situated. This meant that alongside reducing the risk of damage to collections by enemy activity, they also prevented damage by communicable secondary effects such as flood or conflagration from spreading beyond impact zones, enabling proactive risk mitigation measures.[56] Indeed, they lessened the overall threat from aerial attack. But on the other, country houses were, amongst other issues, generally old, made of flammable materials, and often inhabited, rendering them a fire risk.[57] Accordingly, the properties were subject to routine inspections by officials from the Office and later

Ministry of Works, who would check the collections and monitor their conditions. The resulting information ensured that staff at the museum could assess the situation at the refuges, evaluate the merits and demerits of the prevailing strategy, and decide whether any aspect required amending.[58]

Notwithstanding their problems, country houses were generally more conducive to safeguarding the collection of the Imperial War Museum than its own building in Southwark, as later developments would reveal. On 7 September 1940, just over one year after the evacuation, the Luftwaffe launched a sustained bombing campaign against London, known as 'the Blitz'.[59] Throughout the next eight months, the museum became at constant risk from air raids, especially during the night.[60] This was followed by other campaigns, causing forty-one hazardous incidents including a handful of direct hits.[61] The Imperial War Museum was far more vulnerable to falling bombs than other London national museums, such as those in South Kensington. Located near central London, it neighboured multiple potential targets including industry, transport hubs and administrative centres.[62] This meant that despite implementing damage limitation measures at the museum, including fire watches and air raid precautions discussed below, some collection items still sustained irrevocable damage, the impact of which not only degraded the collection, but also the museum as a legitimate institution.[63] Together, the effects forced a review to the existing containment strategy.

The catalyst for change was a direct hit against the Naval Gallery on 31 January 1941 by a high explosive bomb.[64] Alongside expensive ship models and other material, the Imperial War Museum lost the last known remaining Short Seaplane. Flown at the Battle of Jutland during the First World War, this specimen was purportedly the first aeroplane to participate in a naval engagement.[65] Figure 5.2, contrasted with Fig. 3.4, depicts the damage. The Director-General, Bradley, took this incident badly and asked the Trustees to consider his position.[66] But the Board retained confidence in him, with Sir John Forsdyke, Director of the British Museum, writing that 'we [Trustees] are responsible for our policy and decisions in these respects'.[67] Some external observers however were less understanding.

Following the incident, Bradley received a note from the Financial Secretary to the Treasury, Harry Crookshank. Below a document detailing the museum's financial estimates for the following year, he scrawled: 'I am rather shocked to find that this museum has evacuated practically

Fig. 5.2 Photograph depicting the Naval Gallery following a direct hit, c. January-February 1941. © IWM (MH 127)

nothing. Is it too late for them to try and do a bit of dispersing for some of their things are of considerable historical interest?'[68] Crookshank's criticism of the chosen strategy evidences the effect that troubled and poorly received crisis management strategies can have on any museum's image and, in turn, its legitimacy amongst stakeholders.[69] Legitimacy is granted to museums perceived by society as formulating and behaving correctly, however these qualities are defined.[70] The work of crisis containment and limitation therefore involves more than just the work of managing the empirical effects of crisis. It also involves the work of protecting a museum's image in the eye of stakeholders, such as by communicating the effectiveness and appropriateness of the chosen crisis management strategy to the museum.[71] Societal confidence, or the lack thereof, poses significant implications for any organisational recovery from crisis. If crisis management is perceived publicly to be poor, it could catalyse a further and additionally damaging crisis of legitimacy for the impacted museum, as stakeholders and society at large reject the museum with potentially crippling ramifications.[72]

Through the increasing danger and criticism, the Imperial War Museum was forced to change its strategy and evacuate more material, with the Office of Works recommending total evacuation.[73] As previously mentioned however, this was impossible because of the size and weight of some of the collection items. In practice therefore, the institution maintained a partial evacuation, with those collection items which could be removed being removed as and when necessary. Some became accepted by more safely located institutions into their temporary care. Others were sent on tours around the country. And some had space purchased for them by the Office of Works in commercially managed country houses.[74] One of the last elements from the collection to be evacuated was the library. Since war preparations began, it had been intended that library holdings would remain in London as long as possible. This was because they had proven during the inter-war years to be useful resources for undertaking preparatory war research.[75] Eventually, however, that policy became unsustainable. As a result, during June 1941, the library was evacuated to Barnstaple.[76] Following this, practically everything which could be removed from the museum's building, had been.[77]

Recovering from the Trauma

Crises come to an end, eventually. For the museum that has survived, a recovery phase follows. This is stage four of Pearson and Mitroff's five-stage framework. Recovery can be broken down into two parts. The first is a post-crisis assessment that evaluates the needs of the museum to become functional again. This includes determining the schedules and systems required for restoring operations.[78] The second part comprises the execution of a programme of activities which, depending on the findings of the post-crisis assessment, begins restoring the museum's operations and ability to pursue its aims and objectives.[79]

Recovery supports broader efforts at ensuring 'business continuity', what Jamie Watters describes as the process and discipline of 'avoid[ing] any interruption that could lead to either significant losses or failure to achieve the organisations principle objectives'.[80] Similarly to crisis containment and damage limitation, successful recovery depends on extensive and appropriate crisis preparations.[81] A swift recovery is essential for any organisation emerging from surface-defensive crisis. Inactivity from disruption can cause diminished income, market share, public recognition and/or corporate image.[82] Museums must think and operate creatively and entrepreneurially during this stage—such as by performing bricolage, and, where necessary, deploying scaffolding.[83] Accordingly, surface-defensive crisis recovery work can sometimes mimic start-up business operations through being potentially poorly resourced and carried out strategically with small margin for error.[84]

An assessment at the Imperial War Museum by Bradley identified three basic needs to ensure business continuity. The first involved making its building safe and presentable. The second involved receiving and processing new three-dimensional collection items earmarked by the museum for the collection, a matter discussed over subsequent chapters. And the third involved reclaiming the evacuated material, work which continued until 1950.[85] The most important of these was the first, as neither other could be achieved without a usable building.

This matter was initially considered during September 1944 at a meeting convened on the subject by the Ministry of Works with the London national museum directors. It resulted in each museum undertaking an evaluation of the damage they had sustained over the war and determining what work would be needed before fully reopening.[86] By VE-Day, 8 May 1945, the Imperial War Museum had become quite dilapidated. Substantial renovation and cleaning was required before the public could be readmitted. Thus, the museum devised staggered approach to it's restoration, prioritising this work over three stages. The first involved restoring those spaces 'essential for an initial reopening'.[87] The second involved spaces required 'as soon as possible after the initial reopening'.[88] And the third involved spaces which could be reserved 'until complete reopening was possible'.[89] The condition of each space was also analysed and grouped in three categories, with each category classifying the work needed before visitors could enter. The first included spaces requiring 'not much more than cleaning down and/or slight renovation'.[90] The second included spaces requiring 'superficial internal repairs'.[91] And the third included spaces requiring 'light structural repair to the fabric including roofs and/or removal of non-built-in protective treatment'.[92]

From this assessment, it was concluded that the Imperial War Museum could achieve business continuity by reinstating the entrance hall, the reference rooms on Floors B and C in the north block, and the three main galleries on Floor A. The entrance hall was assessed as requiring only minimal work: cleaning down and/or slight renovation. The reference rooms and galleries, by contrast, required superficial and light structural repairs, which did not necessarily prohibit their immediate use.[93] Yet, at the meeting of 27 September 1944, the Ministry of Works warned that through limited resources plus the many homes and critical infrastructure requiring repairs as priority, work on renovating museums needed in the immediate term to be superficial.[94] This involved little more than cleaning and painting, quashing prospects of a prompt permanent reopening once the war had finally ended.

The restrictions placed on renovation put the Imperial War Museum in limbo when VE-Day arrived. Decreasingly needed to facilitate war work, yet unable to resume any public programming until its unoccupied spaces could be restored, the institution looked ahead to an uncertain period of reduced activity and diminishing capital.[95] But in a serendipitous coincidence, that dismal prospect disappeared over June–July 1945 following two chance developments. One was the withdrawal of the *Enemy Weapons Exhibition* from Galleries A and B in July.[96] This externally curated exhibition had resided at the museum since 1944 to familiarise Allied service personnel with enemy weaponry which they might encounter when overseas, the departure of which freed up less dilapidated space for other uses.[97] Another was the opportunity to accommodate a private exhibition by the governmental Petroleum Warfare Department on its wartime work.[98] Bradley had been aware of this exhibition since June 1945 and known that the curators were seeking new venues for public display.[99] After receiving the *Enemy Weapons Exhibition* departure date, he enterprisingly approached the Petroleum Warfare Department and enquired about hosting its exhibition at the museum.[100] The idea was warmly received, so both parties set about collaborating on its transfer.[101]

This was a mutually beneficial arrangement. But it was particularly beneficial for the Imperial War Museum. The exhibition presented a low-cost opportunity to begin restoring normal operations and recommence a public programme following nearly five years of public-facing inactivity. The uptake demonstrated an entrepreneurial spirit required by organisations recovering from crisis: thinking beyond conventional wisdom, quick reactiveness, agility and collaborativeness.[102] Moreover, the museum could gauge its post-war viability through assessing public interest in war—an important development goal of the same institution today.[103] Indeed, the priorities towards developing the visitor base could finally be ascertained. This is essential in maintaining museum sustainability.[104]

The Petroleum Warfare Exhibition opened on 4 October 1945 and remained there until 18 January 1946.[105] 20,856 people attended during that time: an impressive number, considering its only advertisement was through press reviews.[106] This visitor figure, plus the critical acclaim received, led the unpublished war history to treat the exhibition as a success. On 4 March 1946, the museum started preparing for its own permanent if token reopening.[107]

Learning from the Experience

Once a museum has recovered from crisis, they can reflect on their performance. But learning, the fifth and final stage in Pearson and Mitroff's five-stage framework, is frequently neglected.[108] It is also only an emerging area of academic interest.[109] If undertaken after a crisis—particularly surface-defensive crisis, given the ferocity with which they can arise—learning may help museums prepare for the same crisis or similar crises in the future.[110]

As discussed over this chapter and demonstrated by the Imperial War Museum, when crises occur, organisations generally deal with them by following crisis management plans formulated in advance or hurriedly devised in the moment. The organisation responds through taking action to offset or suppress problematic effects and restore operations. This process arises via what Chris Argyris and Donals A. Schön call 'single-loop learning', where 'members of the organization respond to changes in the internal and external environments of the organization by detecting errors which they then correct so as to maintain the central of organizational theory-in-use'.[111] Driven by information and facilitated by evaluation, single-loop learning, a reflexive action, lies at the heart of crisis management activities.

Following the restoration of normal operations, those organisations which do not actively engage in the learning process simply conclude their surface-defensive crisis management actions and put the trauma behind them. The more judicious however extend the crisis management cycle by seeking to understand what just occurred. They also carry out a post-crisis analysis: exploring why the crisis occurred, and how the museum coped with it. Doing so may reveal systemic faults encompassing the underlying cause. It may also reveal better approaches towards implementing future crisis management plans.[112] This process is made possible by what Argyris and Schön call 'double-loop learning' when, after a crisis has been managed, 'organizational inquiry [...] resolve[s] imcompatible [*sic*] organizational norms by setting new priorities and weightings of norms, or by restructuring the norms themselves together with associated strategies and assumptions'.[113] Essentially, where undertaken, the process involves an impacted organisation stabilising the crisis and then modifying its organisational architecture so that the same crisis is less able to resurge.[114]

The geopolitical fallout from the Second World War created a situation in which the Imperial War Museum found itself performing double-loop learning. This transpired soon after VJ-Day when the Ministry of Works convened another conference on 14 December 1945, attended by the London national museum directors, to consider 'whether, having regard for the time in which we lived, we should keep a scheme in being whereby art treasures could be put into safe custody if the need arose'.[115] The Western Allies were now amid the foothills of the Cold War, the new adversary being the Soviet Union.[116] As in 1933, Bradley attended for the Imperial War Museum.[117] The delegates favoured establishing a new scheme, so over the next two years and four months, the Ministry of Works surveyed potential refuges.[118]

At the conference's second meeting on 3 May 1948, the delegates discussed the merits and demerits of the refuges which had been surveyed. Multiple views were heard. Some advocated for below ground refuges, some for above ground refuges, and some for both.[119] Accordingly, the conference chair from the Ministry of Works instructed each delegate to detail in writing their respective institution's requirements. Bradley provided this information on 27 July 1948.[120] The document clearly evidences an output of double-loop learning following the crisis just explored. His twenty-point list, comprising reflections and recommendations for the Imperial War Museum, provides insight into the extent that the institution learned from its Second World War experiences.[121] Through the text, Bradley states what happened, analyses the process, critiques the successes and failures thereof, and states what the museum should do differently to ensure the difficulties encountered would not reoccur. It clearly demonstrates how such an activity can change an organisation's world view so that the organisation develops and applies new beliefs and norms.[122]

According to Bradley, the experience of the Imperial War Museum during the Second World War proved what had been theorised about its premises: that the former-Bethlem Royal Hospital building was extremely vulnerable to aerial attack.[123] But, unexpectedly, the experience also showed that the collection was far less hardy than first appreciated. As per

Bradley's opening statement: 'Experience in the recent war shows that only the toughest objects, guns, tanks, etc., can be left at the museum during wartime with any confidence'.[124] Many more items from the collection had to be evacuated than planned. Thus, Bradley's post-war policy recommendation for storage was that 'In the event of another war we shall aim at almost total evacuation'.[125] Experience also showed the storage of collection items at country houses had been far from ideal. Aside from being difficult to access, difficult to maintain good environmental conditions within, a fire risk, and susceptible to changing hands, Bradley also reflected on their owners' un-altruistic mentality.[126] As he explained: 'while the owners are pleased to receive valuable pictures, they are less willing to receive more bulky but less decorative articles, e.g. ship and other models'.[127] Consequently, Bradley's recommendation for museum refuges was that, in the event country houses again comprised the only refuges available, professionally managed and maintained facilities should be obtained over voluntary ones.[128] This had been arrived at by the museum's experience of using space sublet from the commercial storage company Messrs Bourlet and Sons, which Bradley recalled as having 'aware[ness] for proper conditions and one can feel confident that they are properly seen to'.[129] Alternatively, he recommended that the Ministry of Works oversee the refuges.[130]

The Crisis of Physical Destruction as Surface-Defensive Crisis

Sources of Disruption

The surface-defensive crisis has been introduced above as a crisis that impacts museums physically. This could be by damaging buildings; damaging the things inside buildings, including collections, equipment and other chattels; and/or disrupting staff from performing their duties, amongst other material effects. Having established the premise of the surface-defensive crisis typology, demonstrated how the aerial attacks against London impacted the Imperial War Museum during the Second World War, and demonstrated the way the museum managed this crisis-conducive situation, the chapter now discusses how the bombings, and the institution's response to them, illustrates the surface-defensive crisis typology.

Sources of disruption in a surface-defensive crisis are those things which cause actual tangible harm to museums or prevent them from operating. During peacetime, such sources usually include infrastructural failures such as faulty electricals or plumbing from insufficient building maintenance or damaged roofs from environmental phenomena.[131] Yet, during civil unrest—armed conflict, protests or riots—or natural disaster in unstable countries, likely sources could also encompass violence and theft, including fighting, vandalism and looting.[132] The twentieth and twenty-first centuries offer many examples of museums and other cultural institutions impacted this way beyond the present case study.[133]

The main surface-defensive crisis sources of disruption faced by the Imperial War Museum during the Second World War were the high explosive and incendiary bombs or rockets dropped on or launched at London. This occurred over four phases: the 'Blitz' of 1940–1941, the 'Tip and Run' raids of 1942–1943, the 'Baby Blitz' of 1944, and the V-1 and V-2 strikes of 1944–1945.[134] The intention behind them was to destroy strategic sites and terrorise the population. Recently, evidence has also emerged which reveals how certain London museums and galleries became specific targets of the Luftwaffe.[135] One document in the museum archive, dated 16 September 1940, written by Charles ffoulkes, a Trustee of the Imperial War Museum during the Second World War and the former curator there, also reports that German radio had mentioned the museum as an 'object of attack'.[136]

Potential and Actual Effects

In line with a surface-defensive crisis, these aerial attacks caused significant physical disruption to the Imperial War Museum, most if not all occurring via collateral damage. Many different ordinances were used against London during the Second World War.[137] Typical raids combined high explosive and incendiary bombs to devastating effect.[138] The later V-1 and V-2 strikes were all high explosive.[139] High explosive bombs damaged or demolished buildings and other structures, while incendiary bombs started fires which caused further damage, waylaid the response by local civil defence forces and increased the precariousness of the situation. London Air Raid Precautions warden Barbra Nixon vividly describes the perceived and actual effects of a high explosive bomb against a building:

> Suddenly, before I heard a sound, the shabby, ill-lit, five-storey building ahead of me swelled out like a child's balloon, or like a Walt Disney house having hiccups. I looked at it in astonishment, that bricks and mortar could stretch like rubber. At the point when it must burst, the glass fell out. It did not hurtle, it simply cracked and dropped out, allowing the straining building to deflate and return to normal.[140]

The result was devastation to the building's internal fabric, as occurred with the Imperial War Museum following the strike on the Naval Gallery in January 1941.

The incident with the Naval Gallery was the most destructive event to befall the Imperial War Museum during the aerial attacks. But it was not the only one. Blasts from high explosive bombs detonating nearby repeatedly caused windows to shatter, walls to crack and ceilings to cave, requiring continuous 'first aid repairs' to keep the building functional.[141] This effect was termed 'progressive wrecking' in the institution's unpublished war history, creating unpleasant conditions for the staff who remained there.[142] Other incidents involved incendiary bombs falling on the roof, another dangerous occurrence. These events required prompt responses by staff to prevent the building from catching fire owing to its vulnerable construction: a three wing, four storey angular 'U' shaped structure built of external brick walls; slate, timber and zinc roofing; and internal timber walls and floors.[143] If left unchecked, incendiary bombs could burn ferociously for around 15 minutes.[144]

Its Practical and Critical Significance

The extensive practical significance of the aerial attacks against London to the Imperial War Museum underscores the surface-defensive crisis by the way it prevented the institution from meeting normal aims and objectives. Surface-defensive crises cause disruption through inhibiting organisational capability and capacity.[145] This articulation holds with the classical reading of crisis that something is teetering between continuance and non-existence;[146] an understanding which informs the concept's contemporary medical definition: 'a dangerous state of illness in which it is uncertain whether the sufferer will recover or not'.[147] The aerial attacks on London had just such a practical dimension for the Imperial War Museum. They prevented it from functioning normally and threatened the integrity of the museum's building and its collections, causing certain collection items to be evacuated for their protection and damage to be sustained by the building and those collection items which were not evacuated.

The limited critical (that is, critiquable) significance of the London bombings to the Imperial War Museum also underscores a surface-defensive crisis. Surface-defensive crises derive from ruptures in the lived environment rather than some divergence between institution and society. Hence, a museum impacted by surface-defensive crisis is not one which has departed from societal norms and expectations, but rather become threatened with operational and likely rapid collapse. This restricts the critique of the crisis to the museum itself, for its relationship with wider society is largely irrelevant. Further underscoring a surface-defensive crisis, the Imperial War Museum did not lose any legitimacy from the London bombings. The government continued lending its support to the institution throughout the war. Moreover, recognition that the museum building was a strategic asset ensured any mutterings about the apparent futility of the repairs were kept in check.[148]

Mitigating Measures

The response by the Imperial War Museum to the aerial attacks against London during the Second World War revolved around minimising organisational disruption, the main objective when mitigating surface-defensive crisis. Taking inspiration from medicine, surface-defensive crisis management involves treating the problem through instigating activities that draw on a recovery conceptualisation of resilience. This is the ability to persevere through turbulence in the social system.[149] Put another way, when mid-crisis, the impacted museum must prevent the situation causing its collapse. When post-crisis, the museum must restore as much pre-crisis capability and capacity as possible. The work of surface-defensive crisis management therefore is to maintain an existing paradigm—notwithstanding the need to accommodate structural changes to prevent the crisis' repetition, or the fact that, on a human emotion level, the trauma from crisis can create contexts that feel alien to the beforetimes.[150]

Exemplifying this response, on appreciating the danger of the aerial attacks, the Imperial War Museum pursued a crisis management strategy towards upgrading its defences. The building's capacity to withstand bombing was enhanced, while the staff's capability to mitigate such situations were increased.[151] Accordingly, when the aerial attacks materialised, a reinforced building and staff trained at preventing danger from taking hold ensured the museum survived without irrevocable damage. Moreover, action to evacuate moveable collection items, if first insubstantial, decoupled them from any (further) damage. And once the attacks ceased, the response involved reversing the disruption caused by the bombings.

It is important to note that the staff of the Imperial War Museum were fundamental in resolving this surface-defensive crisis satisfactorily. They conducted fire watching and air raid precautions patrols 24 hours a day, 7 days per week, at great personal risk. Fire watching patrols involved one or two Museum Attendants who would keep watch throughout the building's upper floors for incendiary bombs which might fall on the roof and potentially break through into the internal structure, combusting where they struck.[152] Air raid precautions patrols, by contrast, involved a team of attendants ready to protect locations throughout the institution from any related threat.[153] Through defending the museum in this way, the staff, using their skills, knowledge and experience, prevented the crisis from consuming it. As its unpublished war history states: 'Praise is due to them for having saved the building from fire on three separate occasions without outside assistance'.[154]

Conclusion

The aerial attacks on London severely impacted the Imperial War Museum. They threatened, and sometimes catalysed, considerable physical damage to the museum building and the collection items contained therein, in turn forcing it to close over public safety concerns, which severed contact between the museum and its constituents, and to evacuate collection items from the building. The resulting situation had a significant impact, posing structural, functional, and institutional ramifications that prevented routine operations from being performed during and well beyond the Second World War years themselves.

The Imperial War Museum clearly experienced a dangerous crisis-conducive situation from the various aerial attacks against London. Specifically, the current chapter argues it experienced a surface-defensive crisis, a typology of crisis characterised by tangibly occurring disruption. Through engaging with this concept, the chapter demonstrates that museums are vulnerable to systemic ruptures. Whether deriving internally or externally, from an intentional act or naturally occurring phenomena, they diminish a museum's ability at meeting aims and objectives. The response involves drawing on attributes of resilience to prevent the relevant sources of disruption from arising or, as demonstrated with the response by the Imperial War Museum against the aerial attacks, to prevent them from having some form of catastrophic impact against an institution. Through drawing this conclusion, the chapter argues for robust defensive crisis management strategies when mitigating surface-defensive crises.

The approach of the Imperial War Museum in resolving its crisis of physical destruction was to pursue a strategy which developed and maintained the institution's resilience against aerial attack. After detecting this threat during 1933, the museum prepared various contingency measures, initially alone, but eventually with other bodies and organisations. It identified collection items for evacuation, arranged refuges wherein they could be stored away from harm, and made the collection items remaining *in situ*, plus the museum building generally, as protected and safe as possible. When the aerial attacks eventually began, the measures were implemented. Later, after the Second World War had concluded, the museum undertook a recovery programme that involved restoring the building and instigating a public museological service. And finally, in the wake of this crisis, yet with another, similar and potentially even more destructive crisis looming over the horizon, the museum reflected on its recent experiences and improved the crisis management strategy which had previously been developed. The overall result was the ability to safeguard and continue safeguarding the collection. Without a robust defensive crisis management strategy, the impact posed by the aerial attacks could have been very much worse for the Imperial War Museum. Neither its collection, nor the assets which sustained the collection, may have survived the conflict sufficiently intact, preventing the museum from reopening.

This surface-defensive crisis was not the only crisis typology experienced by the Imperial War Museum during the Second World War. Founded in 1917 as a museum on the 'war to end all war', the Second World War saw it stare down the barrel of incongruence with society once the First World War ceased to represent that utopian idea. This threatened the Imperial War Museum with cultural irrelevancy: an undesirable position for any museum, because it threatens their legitimacy and, in turn, support from stakeholders and constituents. Consequently, the Imperial War Museum undertook concurrent crisis management to avert that possibility from transpiring as well. This comprises the focus of the remaining chapters.

Notes

1. Richard Overy, *The Bombing War: Europe, 1939–1945*, pbk edn (London: Penguin, 2014); Lawrence Ward, *London County Council Damage Maps, 1939–1945* (London: Thames and Hudson, 2015), 112–113.
2. Ernest Blaikley, 'Air Raids and the Imperial War Museum', *Museums Journal* 40 (1941): 297–298.
3. Standing Commission on Museums and Galleries, *Third Report: The War Years and After* (London: HMSO, 1948), 17.
4. Christine M. Pearson and Ian I. Mitroff, 'From Crisis Prone to Crisis Prepared: A Framework for Crisis Management', *The Executive* 7 (1993): 52–53.
5. Edward S. Devlin, *Crisis Management Planning and Execution* (Boca Raton, FL: Auerbach, 2007), 17.
6. E. L. Quarantelli, 'Disaster Research Center', in *Encyclopedia of Sociology*, vol. 1, ed Edgar F. Borgatta and Rhonda J. V. Montgomery, 2nd edn (New York: Macmillan Reference USA, 2000), 681–688.
7. Reinhart Koselleck, 'Crisis', trans. Michaela W. Richter, *Journal of the History of Ideas* 67 (2006): 161.
8. Keith Shaw and Louise Maythorne, 'Managing for Local Resilience: Towards a Strategic Approach', *Public Policy and Administration* 28 (2011): 46.
9. Christer Pursiainen, *The Crisis Management Cycle* (Abingdon: Routledge, 2018), 9–16; Ibrahim Al Luhaidan and Daifallah M. Alrazeeni, 'The Role of Key Early Warning Indicators in Crisis Management: Analytical Review', *Advanced Emergency Medicine* 8 (2019): 9–16.
10. Jan Babecký, Tomáš Havránek, Jakub Matějů, Marek Rusnák, Kateřina Šmídková and Bořek Vašíček, 'Leading Indicators of Crisis Incidence: Evidence from Developed Countries', *European Central Bank Working Paper Series* 1486 (2012): 6–7; Graciela Kaminsky, Saul Lizondo and Carmen Reinhart, 'Leading Indicators of Currency Crisis', *IMF Staff Papers* 45 (1998): 1–48.
11. Vladimir Vukajlović, Ivana Simeunović, Isidora Beraha and Miodrag Brzaković, 'Importance of Information in Crisis Management – Statistical Analysis', *Industrija* 47 (2019): 37–53; Patrick Lagadec, *Preventing Chaos in a Crisis: Strategies for Prevention, Control and Damage Limitation*, trans. Jocelyn M. Phelps (London: McGraw-Hill, 1993), 224–226.
12. Alexandros Paraskevas, 'Mitroff's Five Stages of Crisis Management', in *Encyclopedia of Crisis Management*, ed K. Bradley Penuel, Matt Statler and Ryan Hagen (Thousand Oaks, CA: SAGE, 2013), 628–629; Pursiainen, *Crisis Management Cycle*, 16–23.

13. W. Timothy Coombs, *Ongoing Crisis Communication: Planning, Managing, and Responding*, 5th edn (Thousand Oaks, CA: SAGE, 2019), 33–42.
14. Paraskevas, 'Mitroff's Five Stages', 629; Pursiainen, *Crisis Management Cycle*, 23–26.
15. Brian Milstein, 'Thinking Politically About Crisis: A Pragmatist Perspective', *European Journal of Political Theory*, 14 (2015): 147–149, 151–152.
16. Coombs, *Ongoing Crisis Communication*, 43–44.
17. Christopher Ansell and Martin Bartenberger, *Pragmatism and Political Crisis Management: Principle and Practical Rationality During the Financial Crisis* (Cheltenham: Edward Elgar, 2019), 6.
18. N. H. Gibbs, *Grand Strategy*, vol. 1, *Rearmament*, History of the Second World War (London: HMSO, 1976), 84–85.
19. Terence H. O'Brien, *Civil Defence*, History of the Second World War (London: HMSO, 1955), 12–13.
20. Francisco J. Romero Salvadó, *Historical Dictionary of the Spanish Civil War* (Lanham, MD: Scarecrow, 2013), s.v. 'air force, nationalist' and 'air force, republican'.
21. Stanley Baldwin, *Parliamentary Debates*, Commons, 5th ser., vol. 270 (1932), col. 632.
22. Brett Holman, *The Next War in the Air: Britain's Fear of the Bomber, 1908–1941* (Abingdon: Routledge, 2016), 56.
23. Catherine Pearson, *Museums in the Second World War: Curators, Culture and Change*, ed Suzanne Keene (Abingdon: Routledge, 2017), 63.
24. IWM, Museum Archive (MA), EN2/1/CON/004, 'Safe Custody of National Art Treasures in the Event of War', December 1933; IWM, MA, EN2/1/CON/004, 'Air Raid Precautions: Safe Custody of National Art Treasures', 17 January 1934.
25. Paraskevas, 'Mitroff's Five Stages', 629–630.
26. Ibid.
27. Milstein, 'Thinking Politically', 145.
28. Paraskevas, 'Mitroff's Five Stages', 630.
29. Pursiainen, *Crisis Management Cycle*, 39–95.
30. IWM, MA, EN2/1/CON/004, 'Safe Custody of National Art Treasures in the Event of War', December 1933; IWM, MA, EN2/1/CON/004, 'Air Raid Precautions: Safe Custody of National Art Treasures', 17 January 1934.
31. Pursiainen, *Crisis Management Cycle*, 85–88.
32. Stephanie Duchek, 'Organizational Resilience: A Capability-based Conceptualization', *Business Research* 13 (2020): 226.

33. Steven Cooke and Lloyd Jenkins, 'Discourses of Regeneration in Early Twentieth-Century Britain: From Bedlam to the Imperial War Museum', *Area* 33 (2001): 382–390; Pursiainen, *Crisis Management Cycle*, 79–81.
34. IWM, MA, EN1/1/REP/032, 'War History of the Imperial War Museum, 1933–1943', 4.
35. IWM, MA, EN2/1/MUS/002/16, de Normann to Curator and Secretary, 24 September 1937.
36. IWM, MA, EN1/1/REP/032, 'War History […], 1933–1943', 1
37. Committee of Public Accounts, *Special Report and First, Second and Third Reports from the Committee of Public Accounts Together with the Proceedings of the Committee, Minutes of Evidence, Appendices and Index* (London: HMSO, 1957), 17.
38. IWM, MA, EN2/1/MUS/002/16, de Normann to Curator and Secretary, 24 September 1937.
39. Paraskevas, 'Mitroff's Five Stages', 629.
40. Ibid.
41. William 'Rick' Crandall, John A. Parnell and John E. Spillan, *Crisis Management: Leading in the New Strategy Landscape*, self-pub. 4th edn (n.place: n.pub., 2021), 204, note that this monograph was formally published by SAGE in 2014.
42. Paraskevas, 'Mitroff's Five Stages', 630.
43. George R. Franks, 'Damage Containment', in *Encyclopedia of Crisis Management*, ed K. Bradley Penuel, Matt Statler and Ryan Hagen (Thousand Oaks, CA: SAGE, 2013), 226–230.
44. Niall Ferguson, *Doom: The Politics of Catastrophe* (London: Allen Lane, 2021).
45. Paul Hopkin, *Fundamentals of Risk Management: Understanding, Evaluating and Implementing Effective Risk Management*, 4th edn (London: Kogan Page, 2017), 156.
46. Coombs, *Ongoing Crisis Communication*, 52–53.
47. Jennifer Wilson, 'Now the Disaster's Happened, What Am I Supposed to Do?', Accounting Today, https://www.accountingtoday.com/news/now-the-disasters-happened-what-am-i-supposed-to-do (18 March 2024).
48. *Crisis Management* (Boston, MA: Harvard Business School Press, 2004), 66–68.
49. Pearson, *Museums*, 64.
50. Alexandros Paraskevas, 'Risk Treatment', in *Encyclopedia of Crisis Management*, ed K. Bradley Penuel, Matt Statler and Ryan Hagen (Thousand Oaks, CA: SAGE, 2013), 842.

51. IWM, MA, EN2/1/MUS/002/3, six-page untitled and undated document headed with 'Class I' on page 1, 'Class II' on page 4, 'Class III' on page 5, and 'Class IV' on page 6; IWM, MA, EN2/1/MUS/002/3, 'Photographic Albums'.
52. IWM, MA, EN1/1/REP/032, 'War History [...], 1933–1943', 4.
53. IWM, MA, EN2/1/MUS/002/3, six-page untitled and undated document headed with 'Class I' on page 1, 'Class II' on page 4, 'Class III' on page 5, and 'Class IV' on page 6.
54. IWM, MA, EN1/1/REP/032, 'War History [...], 1933–1943', 4–5.
55. IWM, MA, EN2/1/MUS/002/3, one-page untitled and undated document detailing the name and location of the refuges and contact details of their owners.
56. Paraskevas, 'Risk Treatment', 842–843; Franks, 'Damage Containment', 227.
57. IWM, MA, EN2/1/GOV/111, 'Storage Accommodation Required by the Imperial War Museum in the Event of War', 27 July 1948; McCamley, *Saving Britain's Art Treasures* (Barnsley: Leo Cooper, 2003); Debbie Whittaker, 'Museums at War: The Impact of the Second World War on the National Museums of London, with Particular Reference to the Imperial War Museum, National Gallery, Victoria and Albert Museum and the British Museum' (master's dissertation, University of Leicester, 2010); John Martin Robinson, *Requisitioned: The British Country House in the Second World War* (London: Arum, 2014), 1–2.
58. The National Archives (TNA), WORK 17/180, see documents therein; Crandall, Parnell and Spillan, *Crisis Management*, 216–217.
59. Overy, *The Bombing War*, 86.
60. IWM, MA, EN2/1/MUS/002/2, see documents contained therein. Note that this documentation does not provide full details of the effects on the Imperial War Museum, but it does covey a sense as to the experience.
61. IWM, MA, EN1/1/REP/032, 'War History of the Imperial War Museum: Part II, January 1944 to the Reopening of the Museum on 27th November 1946', 4. Note that the source of this information states 10 September 1941–14 February 1945, which appears to be a typographical error, as the first hazardous incident threatening the Imperial War Museum is documented in EN2/1/MUS/002/2 and elsewhere as taking place on 9 September 1940, forcing the institution's closure on 10 September until 4 October 1945.
62. Ward, *Bomb Damage Maps*, 108–109, 112–113; Caroline Shenton, *National Treasures: Saving the Nation's Art in World War II*, pbk edn (London: John Murray, 2022), 170.
63. Hopkin, *Fundamentals*, 156.

64. IWM, MA, EN2/1/ACC/004/7, Bradley to Cubitt, 1 February 1941.
65. Ernest Blaikley, 'Reopening of the Imperial War Museum', *Museums Journal* 40 (1940): 8.
66. IWM, MA, EN2/1/ACC/004/7, Bradley to Cubitt, 1 February 1941.
67. IWM, MA, EN2/1/ACC/004/7, Forsdyke to Bradley, 5 February 1941.
68. IWM, MA, EN2/1/COB/001/3, Standing Committee, 20 February 1941, 1.
69. Joseph Eric Massey, 'Managing Organizational Legitimacy: Communication Strategies for Organizations in Crisis', *The Journal of Business Communication* 38 (2001): 153–183.
70. Mark C. Suchman, 'Managing Legitimacy: Strategic and Institutional Approaches', *Academy of Management Review* 20 (1995): 574–575.
71. Martin N. Ndlela, *Crisis Communication: A Stakeholder Approach* (Cham: Springer Nature, 2019), 99–104; Jasmine N. Duran, 'Headliners: The Influence of Media Coverage in Museum Crises' (unpublished master's thesis, The University of the Arts, 2014).
72. Finn Frandsen and Winni Johansen, 'Crisis Communication and Organizational Legitimacy', in *Handbook of Business Legitimacy: Responsibility, Ethics and Society*, ed Jacob Dahl Rendtorff (Cham: Springer Nature, 2020), 707–725.
73. IWM, MA, EN2/1/MUS/002/2, three-page undated document headed with 'Imperial War Museum' on page 1 and subheaded with 'Air Raid Incidents' on pages 1–2, 'Evacuations', 'Loans' and 'Disposals' on page 2 and 'Miscellaneous Happenings' on page 3, 2.
74. IWM, MA, EN1/1/REP/032, 'War History [...], 1933–1943', 6–7.
75. IWM, MA, EN1/1/REP/032, 'Imperial War Museum: 21st Annual Report of the Director-General to the Board of Trustees', 1.
76. IWM, MA, EN1/1/REP/032, 'War History [...], 1933–1943', 7.
77. See Philip W. Deans, 'Protective Measures Before the 1954 Hague Convention: The Case of the Imperial War Museum, 1933–1950', in *Safeguarding Cultural property and the 1954 Hague Convention: All Possible Steps*, ed Emma Cunliffe and Paul Fox (Woodbridge: Boydell, 2022), 57–76.
78. John Child, *Organization: Contemporary Principles and Practice* (Malden, MA: Blackwell, 2005), 12; Paraskevas, 'Mitroff's Five Stages', 630–631.
79. Paraskevas, 'Mitroff's Five Stages', 631.
80. Jamie Watters, *The Business Continuity Management Desk Reference: Guide to Business Continuity Planning, Crisis Management and IT Disaster Recovery* (n.place: Business Leverage, 2010), 9.
81. Paraskevas, 'Mitroff's Five Stages', 631.

82. Yunus D. Saleh, *Crisis Management: The Art of Success and Failure—30 Case Studies in Business and Politics* (Minneapolis, MN: Mill City, 2016), 258.
83. James M. Kendra and Wachtendorf, 'Elements of Resilience After the World Trade Center Disaster: Reconstituting New York City's Emergency Operations Centre', *Disasters* 27 (2003): 42; Wanda J. Orlikowski, 'Material Knowing: The Scaffolding of Human Knowledgeability', *European Journal of Information Systems* 15 (2006): 461–462.
84. Bianca Miller-Cole and Byron Cole, *Self Made: The Definitive Guide to Business Startup Success* (London: John Murray Learning, 2017).
85. IWM, MA, EN2/1/GOV/103, 'Notes on Immediate Post-War Policy and Storage Requirements. For Discussion at the Meeting on May 1st'; IWM, MA, EN2/1/COB/009/2, Board of Trustees, 27 March 1950, 3.
86. IWM, MA, EN2/1/GOV/103, 'Museums and Galleries', 27 September 1944.
87. IWM, MA, EN2/1/GOV/103, 'Art and Science Building Post-War Reopening'.
88. Ibid.
89. Ibid.
90. Ibid.
91. Ibid.
92. Ibid.
93. Ibid.
94. IWM, MA, EN2/1/GOV/103, 'Museums and Galleries', 27 September 1944.
95. Saleh, *Crisis Management*, 258.
96. IWM, MA, EN2/1/GOV/058/1, anon. to Robson, 12 July 1945.
97. IWM, MA, EN1/1/REP/032, War History [...]: Part II', 4–7.
98. IWM, MA, EN2/1/GOV/058/1, anon. to Robson, 12 July 1945.
99. Ibid.
100. Ibid.
101. IWM, MA, EN2/1/GOV/058/1, Robson to Bradley, 21 July 1945; IWM, MA, EN2/1/GOV/058/1, 'P.W.D. Exhibition at the Imperial War Museum', 25 July 1945; IWM, MA, EN2/1/GOV/058/1, 'Petroleum Warfare Department Exhibition, Imperial War Museum', 11 September 1945.
102. Paraskevas, 'Mitroff's Five Stages', 631; Dan Sullivan, 'The Power of Entrepreneurship During A Crisis', Multiplier Mindset Blog, Strategic Coach, https://resources.strategiccoach.com/the-multiplier-mindset-blog/the-power-of-entrepreneurship-during-a-crisis (18 March 2024).
103. Stephen Moss, 'The Imperial War Museum: As Much a Relic as Its Spitfires and Doodlebugs?', *Guardian*, 10 July 2014, para. 2, https://

www.theguardian.com/culture/2014/jul/10/imperial-war-museum-reopens-first-world-war (18 March 2024).
104. John Reeve, 'Prioritizing Audience Groups', in *The Responsive Museum: Working with Audiences in the Twenty-First Century*, ed Caroline Lang, John Reeve and Vicky Woolard (Aldershot: Ashgate, 2006), 43–60.
105. IWM, MA, EN2/1/GOV/058/1, sample invitation to exhibition opening; IWM, MA, EN2/1/COB/005/2, Board of Trustees, 4 March 1945.
106. IWM, MA, EN1/1/REP/032, War History [...]: Part II', 19.
107. IWM, MA, EN2/1/COB/005/2, Board of Trustees, 4 March 1945.
108. Mark D. Cannon and Amy C. Edmondson, 'Confronting Failure: Antecedents and Consequences of Shared Beliefs About Failure in Organizational Work Groups', *Journal of Organizational Behaviour* 22 (2001): 265; Crandall, Parnell and Spillan, *Crisis Management*, 265.
109. Pursiainen, *Crisis Management Cycle*, 146.
110. Abraham Carmeli and John Schaubroeck, 'Organisational Crisis-Preparedness: The Importance of Learning from Failures', *Long Range Planning* 41 (2008): 177–196.
111. Chris Argyris and Donals A. Schön, *Organizational Learning: A Theory of Action Perspective* (Reading, MA: Addison-Wesley, 1978), 18.
112. Paraskevas, 'Mitroff's Five Stages', 631.
113. Argyris and Schön, *Organizational Learning*, 24.
114. Pursiainen, *Crisis Management Cycle*, 152.
115. IWM, MA, EN2/1/GOV/097, 'Storage of National Art Treasures', 14 December 1945, 1.
116. Adewunmi J. Falode and Moses J. Yakubu, 'A New Historiography of the Origins of the Cold War', *Jurnal Sosial dan Humaniora* 9 (2019): 101–104.
117. IWM, MA, EN2/1/GOV/097, 'Storage of National Art Treasures', 14 December 1945, 1.
118. Ibid.
119. IWM, MA, EN2/1/GOV/111, 'Storage of National Art Treasures', 3 May 1948.
120. IWM, MA, EN2/1/GOV/111, 'Storage Accommodation Required by the Imperial War Museum in the Event of War', 27 July 1948.
121. David A. Kolb, *Experiential Learning: Experience as the Source of Learning and Development* (Upper Saddler River, NJ: Pearson Education, 2014).
122. Eleanor Stead and Clive Smallman, 'Understanding Business Failure: Learning and Un-learning Lessons from Industrial Crises', *Understanding Business Failure* 7 (1999): 5.
123. IWM, MA, EN2/1/GOV/111, 'Storage Accommodation Required by the Imperial War Museum in the Event of War', 27 July 1948, 1.
124. Ibid.

125. Ibid.
126. Ibid., 2.
127. Ibid.
128. Ibid.
129. Ibid.
130. Ibid.
131. Barrie Chanter and Peter Swallow, *Building Maintenance Management*, 2nd edn (Oxford: Blackwell, 2007), 133; John R. Brumgardt, 'Mitigating Staff Stress in a Natural Disaster: The Charleston Museum, Charleston, S.C.', in *Institutional Trauma: Major Change in Museums and Its Effect on Staff*, ed Elaine Heumann Gurian (Washington, DC: American Association of Museums, 1995), 59–74.
132. Laurie R. Blank and Gregory P. Noone, *International Law and Armed Conflict: Fundamental Principles and Contemporary Challenges in the Law of War* (New York, NY: Wolters Kluwer Law and Business, 2013), 345; Philip Nel and Marjolein Righarts, 'Natural Disasters and the Risk of Violent Civil Conflict', *International Studies Quarterly* 52 (2008): 159–185.
133. Robert Bevan, *The Destruction of Memory: Architecture at War*, 2nd edn (London: Reaktion, 2016).
134. Overy, *The Bombing War*, 73, 120; Chris Goss, with Peter Cornwell and Bernd Rauchbach, *Luftwaffe Fighter-Bombers Over Britain: The Tip and Run Campaign, 1942–1943* (Mechanicsburg, PA: Stackpole, 2003); Christie Campbell, *Target London: Under Attack from the V-weapons During WWII* (London: Abacus, 2013).
135. Sam Tonkin, 'Hitler's Plans to Destroy London: Rare Mare Map Revealing Germany's WWII Bombing Targets Is Discovered After 75 Years in An Attic', *Mail Online*, 24 February 2017, https://www.dailymail.co.uk/news/article-4256244/Map-revealing-Germany-s-WWII-bombing-targets-discovered.html (18 March 2024).
136. IWM, MA, EN2/1/COB/039, 'Report of Air Raid Damage'.
137. John Ray, *The Night Blitz, 1940–1941*, new edn (London: Cassell Military, 2000), 83–86.
138. Overy, *The Bombing War*, 76.
139. Steven J. Zaloga, *V-1 Flying Bomb, 1942–52* (Botley: Osprey, 2005); Steven J. Zaloga, *V-2 Ballistic Missile, 1942–52* (Botley: Osprey, 2003).
140. Barbra Nixon, *Raiders Overhead: A Diary of the London Blitz*, ed Ian Grant (London: Scolar, 1980), 26.
141. IWM, MA, EN1/1/REP/032, 'War History [...], 1933–1943', 14.
142. Ibid., 10.
143. Ibid., pp. 3, 10, 13.
144. Overy, *The Bombing War*, 76.

145. Pursiainen, *Crisis Management Cycle*, 79–81, 85–88.
146. Koselleck, 'Crisis', 161.
147. Harvey Marcovitch, *Black's Medical Dictionary*, 43rd edn (London: Bloomsbury Information, 2017), s.v. 'crisis'.
148. IWM, MA, EN1/1/REP/032, 'War History [...], 1933–1943', 19–20.
149. Brigit Maguire and Sophie Cartwright, *Assessing a Community's Capacity to Manage Change: A Resilience Approach to Social Assessment* (Canberra, ACT: Bureau of Rural Science, 2008), 4, https://www.agriculture.gov.au/sites/default/files/abares/documents/dewha-resilience-sa-report-final-4.pdf (18 March 2024).
150. Rina Bar-Lev Elieli and Laurence J. Gould, 'Introduction', in *Institutional Trauma: Major Change in Museums and its Effect on Staff*, ed Elaine Heumann Gurian (Washington, DC: American Association of Museums, 1995), 29.
151. Pursiainen, *Crisis Management Cycle*, 79–81, 85–88.
152. IWM, MA, EN2/1/STA/001, 'Fireguarding', 20 October 1942.
153. IWM, MA, EN2/1/MUS/002/5, three-page untitled and undated document headed with 'A.R.P. Patrol' and 'The Function of A.R.P. Patrol' on page 1, 'Equipment', 'Gas Alarm', 'Prolonged Spells of Duty' and 'Stirrup Pumps and Redhill Units' on page 2, and 'Duty Period', 'Staff Shelter', 'Alternative Use of Staff Shelter as A.R.P. Headquarters' and 'General Orders' on page 3.
154. IWM, MA, EN1/1/REP/032, 'War History [...], 1933–1943', 19.

CHAPTER 6

Reconceiving the Institution, 1939

Introduction

Crises have been defined in this study as uncomfortable, unpredictable, unstable and potentially dangerous situations, where the impacted museum will be disrupted, perhaps inoperably and irreparably, requiring extraordinary intervention to be overcome. By describing them as disruptive however, this definition encompasses more than just physical, dangerous disruption that threaten the operations of museum or the life and limb of their staff, as potentialised by surface-defensive crises explored in Chap. 5. It also encompasses non-physical disruption, including shifting societal values. Hence, different crisis conditions cause different situations, requiring different resolutions and strategies, as appropriate.

Following on from Chaps. 4 and 5, this study continues exploring ways in which the Imperial War Museum was impacted by and responded to crisis-conducive situations during the Second World War era. Specifically, over the next three chapters, it explores a perceived threat from evolving societal interests, behaviours and practices of the visiting public immediately following the Second World War, a phenomenon Montgomery Van Wart terms 'value shifts'.[1] The following chapters argue that museums which fail to make sufficient corresponding 'value adjustments' risk encountering diminishing legitimacy and sustainability. Indeed, when overcoming such crises, they must mount a revolutionary response, which corresponds with the modernity-esque reading of the concept discussed in

P. W. Deans, *Crisis, Reinvention and Resilience in Museums*,
https://doi.org/10.1007/978-3-031-67806-6_6

Chap. 2. This is where the museum 'emancipates' themselves from their crisis conditions by organisationally developing themselves such that the crisis conditions are diminished.[2]

The current chapter contributes towards this part of the study by exploring the response of the Imperial War Museum to an anticipated crisis of irrelevancy as the Second World War loomed and took hold. Along with Chaps. 7 and 8, it mounts an argument that this threat caused a crisis-conducive situation which was only resolvable by the museum undergoing institutional reinvention. In exemplifying this necessary revolutionary response for dealing with the situation, the chapter also independently discusses how such revolutionary responses require socio-political regulatory legitimacy to succeed.

The historical context of the current chapter is the busy and often anxiety-ridden period spanning September–December 1939, when the United Kingdom was preparing for and adjusting to fighting another war against Germany. Thitherto, concern that the Imperial War Museum might lose cultural relevancy if the institution did not update its subject remit towards representing any future war experiences had arisen and become deeply embedded there, particularly with the Director-General, Leslie Bradley. He wrote very pessimistically about this fear following the war's onset. In one example, on 12 September 1939, Bradley informed Henry Ludwig Mond, 2nd Baron Melchett, a Trustee of the Imperial War Museum, his view that 'the extension of the Museum [...] seems to me to be absolutely vital to its continued existence'.[3] And in another, on 22 September 1939, he informed the Treasury Secretary his belief that, were the changes not to take place, 'the interest and utility of the museum will be greatly reduced if, and when, it is reopened to the public'.[4]

To understand the crisis of irrelevancy that threatened the Imperial War Museum and explore the institution's response against it, the current chapter draws on two crisis concepts. One is the 'deep-revolutionary crisis', the second of two novel crisis typologies conceived and theorised towards this study, established below, which characterises the crisis-conducive situation experienced by the museum. The other concept is Christine M. Pearson and Ian I. Mitroff's general five-stage framework for crisis management, formally introduced in Chap. 2 and used throughout Chap. 5.[5] The chapter also draws on the reinvention formula proposed by David Gleicher and shared by Richard Beckhard, again formally introduced in Chap. 2, to appreciate the structure and nature of the museum's response.[6]

The current chapter addresses the above over three substantive sections. The first section introduces and explicates deep-revolutionary crisis.

It sets out the typology's premise and signification. The second then explores the initial response by the Imperial War Museum to the threatened and feared value shift. It explores the sense making and preparatory work which the museum carried out, drawing on the first two stages of Pearson and Mitroff's five-stage crisis management framework: 'signal detection' and 'preparation/prevention'. And the third explores the received wisdom surrounding the Director-General, Leslie Bradley, particularly his legacy at the Imperial War Museum. It assesses the accuracy and fairness of that received wisdom, reflecting on the role which Bradley played in mitigating this crisis. Finally, it should be noted that, because the response to the threatened value shift is considered over three chapters, the current chapter does not incorporate any discussion on how the case exemplifies a deep-revolutionary crisis, as was included with Chap. 5 *apropos* surface-defensive crisis. This discussion takes place in Chap. 8, ensuring the case can be fully explicated before any resemblances are drawn.

Deep-Revolutionary Crisis

The deep-revolutionary crisis is another organisational crisis typology predicated on two descriptors that contrasts with the surface-defensive crisis introduced in Chap. 5. The first descriptor, 'deep', establishes the practical and critical significances of all applicable crisis situations as well as the implications which these crises have for an organisation. It highlights the immaterial, that is the non-physical, dimension which they will manifest in and cause impact to organisations. It also conveys their point of focus as an analytical tool, these being on organisational matters beneath the surface-level, practical, everyday organisational operations. The second descriptor, 'revolutionary', establishes the necessary organisational response in dealing with all applicable crises. It references the radical posture of the mitigating measures that should be deployed by organisations against them. It also signifies the various dangers which organisations may face on being impacted. These can include anything leading to a schism or incongruency between an organisation and their clientele, such as disapproval with the organisation's actions or inaction over something, or divergence in the organisation's role or purpose and prevailing expectations thereof, respectively.[7]

A deep-revolutionary crisis may at first appear innocuous compared with a surface-defensive crisis. But maintaining this view would be to misapprehend the significance of the typology. They can profoundly and insidiously impact a museum. While the deep-revolutionary crisis may not cause physical, visually perceptible disruption, it can destabilise museum *raisons d'être* and rationales, potentially over protracted periods. In doing so, the typology disrupts the legitimacy sustaining a museum. Accordingly, while they pose little if any danger to the functional integrity of a museum, they are, nonetheless, the reason why operationally sound organisations collapse: through losing the necessary social, political, economic and cultural support to remain viable over the mid- and long-term.

The aim of deep-revolutionary crisis management is to replace a failing organisational paradigm with an alternative, more sustainable model. As Fig. 6.1 illustrates, this involves making the necessary changes to an organisation until the issues causing the misalignment are absorbed by it, neutralising their corrosive effects. Otherwise, the crisis risks eating away at the museum's legitimacy until it can no longer command the necessary social, political and financial support from stakeholders and wider society. Deep-revolutionary crisis management is therefore indicative of a transitional process: a pushing forward through the crisis to a fresh coherence with the social system and a new social contract.[8] Consequently, the deep-revolutionary crisis draws on the concept's modernity-esque reading.[9] Moreover, it overlaps with the transformation conceptualisation of

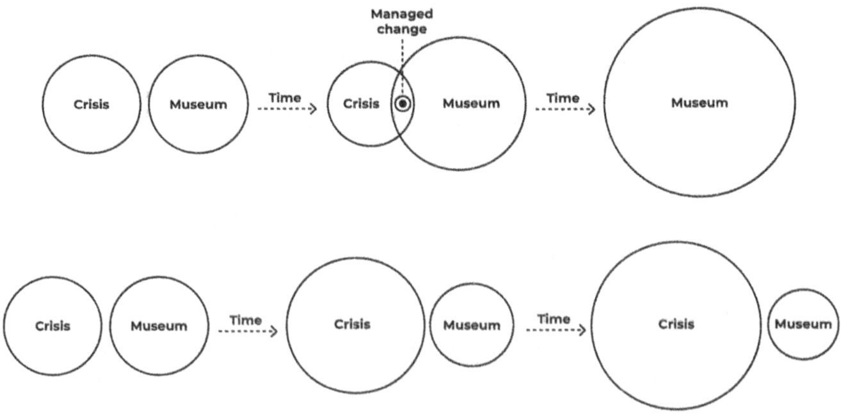

Fig. 6.1 Diagram illustrating two possible outcomes of deep-revolutionary crisis. Note that the museum which accepts and takes on (absorbs) the issues causing its misalignment with the social system preserves or increases its legitimacy, whereas the museum which fails to take on the issues experiences diminishing legitimacy as the crisis develops

organisational resilience discussed in Chap. 4, where recourse to resilience involves adapting away from difficulty.[10]

Resolving a Crisis of Irrelevancy at the Imperial War Museum (I): Determining the Response

Sensing the Threat

As discussed in Chaps. 2 and 5, one starting point of museum crisis management are activities undertaken towards signal detection.[11] This involves scanning the environment for suspect signals and assessing them to distinguish red flags from background noise and to ascertain their credibility.[12] It is indeterminable when the Imperial War Museum began assessing the risks of a potential societal value shift in the wake of another armed conflict or the threat posed by post-war cultural irrelevancy. The same can also be said for the exact person who first recognised and appreciated the problem. This transpires because the earliest discovered document conveying first-hand concerns over the matter offers no insight into how long the threat had been conceived for, nor by whom. The document in question is a letter dated 7 June 1939 by Leslie Bradley, Director-General of the Imperial War Museum, to P. F. R. Beards, the Private Secretary of the Deputy Under-Secretary of State for War:

> I have from time to time urged upon the Chairman of the Board, Sir Bertram Cubitt, the desirability of establishing relations with the Service, and other Departments, with a view to the preservation of the records and relics of any future war, but his attitude is, first, that such Departments have far too many other things to consider, and that in any case the handing over of such relics and records to this Museum will be automatic in the event of another war.[13]

What can be determined however is that red flags had been detected and disseminated at the Imperial War Museum well before the Second World War commenced. Moreover, it reveals that Bradley himself had received the signals, legitimised them, and was well into contemplating the necessary resolution for the problem: a new contemporary collecting programme, which would enable the Imperial War Museum to, in effect, reinvent itself.[14]

Another informative aspect of this letter is the insight it offers into the way Bradley was making sense of the prospective danger of post-war cultural irrelevancy. Sense making plays an important role in signal detection, indeed crisis management generally. The earlier an emerging threat can be identified, the sooner it can be dealt with. Moreover, if during this process the sources, traits and ramifications of a crisis are quickly understood, the more likely the crisis will be mitigated.[15] But sense making becomes especially important for deep-revolutionary crisis management. Unlike surface-defensive crises discussed in Chap. 5, deep-revolutionary crises do not overtly exhibit danger. This means that their practical and critical significance may be more elusive than crises of empirical origin.

Despite its importance to the efficacy of crisis management, sense making, as with signal detection overall, is not easily accomplishable. A museum's complexity may mask a crisis' onset. Structural barriers may prevent staff from raising the alarm or mitigating infrastructure from being established. And the subjectiveness of crisis may mean that some do not perceive any threat where one exists, or *vice versa*.[16] Respondents to deep-revolutionary crises, therefore, must be adept at reading crisis situations, otherwise they risk misappreciating them.[17] While no specific attributes suffice sense making, research shows that experience and intensive training, intuition, and organisational structures or cultures that institute proactive scanning for problems and threats, can all help facilitate the process.[18]

Bradley's emergent sensing of the sources, characteristics and consequences of the potential societal value shift confronting the Imperial War Museum in the wake of the Second World War is demonstrated through his determination that the institution should not suffer from the same handicap which hampered its collecting activities during the previous world war. When the War Cabinet instituted the Imperial War Museum on 5 March 1917, the First World War had been underway for more than two years.[19] Over this time, a considerable amount of war material, particularly ephemera, entered circulation, but soon disappeared without any specimens having ever been collected. During many years to come therefore, numerous pre-1917 items, specimens big and small, would remain unrepresented in the museum's collection. A repeated experience threatened the same implications for any new collection. This, in turn, risked limiting the effectiveness of the crisis management strategy. After all, the more fulsome the new contemporary collection became, the more relevant the museum should be to future audiences and researchers' wants and needs.[20]

This concern over losing the opportunity to collect material from the new ongoing conflict is clearly evidenced through three letters which

Bradley sent various senior governmental figures during September 1939. In each, he earnestly and urgently requested that the recipient assisted the museum with a contemporary collecting initiative by organising, or at least influencing, the informal earmarking and storing of material across their respective departments until the museum could assume custody. The first such letter was sent on 11 September 1939 to the Secretary of State for Air, Sir Kingsley Wood:

> I consider it my duty to bring this matter to your notice now because the utility of the Museum and the completeness of its record have always been hampered by the fact that the systematic collection of material did not begin until 1917, and unless immediate steps are taken in this matter another unrecoverable gap will be formed. [...] How long, for instance, will it be possible to obtain copies of the leaflets which have been dropped in Germany?[21]

The second was sent to the Lord Privy Seal, Sir Samuel Hoare, a day later:

> [The Imperial War Museum] has never been able to attain completeness owing to the impossibility of tracing all that was lost during the first half of the [last] war when no systematic effort was made to collect a record of activities on what is commonly called the 'Home Front'. [...] Our experience, and that of the many who have made use of our records for the completion of books about the Great War, is that the very beginning of a war is the only time for such collection to be begun, a final pruning being much simpler than the filling of complete gaps.[22]

And the third was sent to the Home Secretary, Sir John Anderson, who also doubled up as Minister for Home Security, a further two days on from the second:

> Both we and they [those who used the museum's collections for war research during the run up to the Second World War] have [...] suffered from comparative paucity of records and material for the earlier half of the [last] war when no systematic collection was made, much of the material disappearing, for instance, when the paper shortage led to the pulping down of all obsolete printed papers, and in the constant stress of dealing immediately with fresh problems. [...] [T]he last few years have amply proved the convenience of keeping a complete record of war measures accessible for reference in a future emergency.[23]

The ramifications of not extending the subject remit were also underscored in a letter which Bradley sent to Sir James Ross, Deputy Secretary of the Air Ministry and Trustee of the Imperial War Museum, on 24 October 1939. Summed up in one sentence, he wrote: 'I could see no future for the Museum if it was to remain merely a Museum of the last war but one'.[24]

Although Bradley cannot be confirmed as the individual who first detected the threat of post-war cultural irrelevancy at the Imperial War Museum, he had certainly become convinced by the danger it posed and sensed that an appropriate strategy involved updating the institution's *raison d'être* and rationale. Through reaching this understanding, the four core requirements for major systemic change established in Chap. 2 were being attained by the museum: dissatisfaction with the *status quo*, an emerging conception of an alternative desired state, understanding about the necessary first steps for reaching that state, and an imperative to undergo change which outweighed the process's overall holistic cost.[25]

Seeking Permission to Respond

After detecting and legitimising crisis signals, a museum, now operating under crisis conditions, must prepare to mitigate the incoming crisis—including, where possible, working to prevent it from striking.[26] The appropriateness of the course depends on the nature of the crisis being confronted.[27] For example, the threats posed by aerial attacks on London, discussed in Chap. 5, could not be prevented from impacting the Imperial War Museum. There, the museum was required to take the former course and prepare to resist the impact, an approach consistent with surface-defensive crisis management. This involved building up its organisational resilience ahead of any armed conflict which might occur. Conversely, the threat posed by a potential societal value shift following another armed conflict could be prevented from impacting the Imperial War Museum. Here, the museum was required to take the latter course and work as outlined below to prevent the impact from materialising in the first place, an approach consistent with deep-revolutionary crisis management.

By summer 1939, Bradly had become convinced by the necessity of extending the subject remit on which the Imperial War Museum had been founded. But conviction alone was not enough to instigate the process. He needed permission from the Board of Trustees—those who held ultimate responsibility for the museum and who would be liable if the extension caused problematic repercussions. Hence, the prevention stage of this crisis-conducive situation can be seen as unfolding over a twofold process. The first, considered in the current chapter, involved seeking the requisite permission to make the necessary changes. The second part, which the following two chapters focus on, involved executing the change.

This need for approval by the Board of Trustees counters a misconception that crisis strategy and decision making is the responsibility of executive leaders and those leaders alone. On the contrary, such circumstances are rarely the case. Usually, the formulation and execution of crisis management involves multiple personnel with varying seniority and skill sets, including board members.[28] The same applies to organisational change. Museums cannot be changed by their leaders in a vacuum or silos. The process requires input from agents at all echelons and departments to ensure it is not only effective but also appropriate.[29]

Bradley formally raised the extension of the subject remit with the Board of Trustees on 20 June 1939.[30] Although this discussion was the first minuted on the topic since at least November 1933, it would unlikely have been the Trustees' first encounter with the idea over that period. As suggested by his letter to P. F. R. Beards on 7 June 1939, Bradley shows himself a vocal proponent of the extension. Moreover, the way that the minutes record how the matter was raised implies it had previously been discussed off the record. For example, they stated: 'The Secretary reported that in connection with the consideration of the question of the preservation and custody of relics in the event of war [...]'.[31] Yet, if Bradley had hoped for a quick approval, he may have been disappointed by what followed. In response, the Trustees 'laid it down that it was for the Government to decide whether and where such relics and records were to be kept'.[32] The reason for this decision stems from two concerns.

The first concern revolved around the regulative legitimacy of the extension, relative to the provisions made by the *Imperial War Museum Act* 1920—the primary legislation which formally incorporated the Imperial War Museum nineteen years previously. When Bradley proposed his plan on 20 June 1939, he did so while acknowledging that certain individuals at the War Office had queried whether the scheme was fundamentally lawful.[33] It seems this admission raised serious concerns amongst the Trustees over the socio-political regulative legitimacy of what Bradley was proposing.[34] Unlike surface-defensive crisis management which focusses on preserving organisational paradigms, deep-revolutionary crisis management is focussed on creating new ones. Such action needs to remain in alignment with the prevailing frameworks that govern a museum, otherwise it risks attracting sanctions and disrepute.[35] This outcome can catalyse a broad reduction of legitimacy as discussed in Chap. 5, significantly reducing support from stakeholders and society at large.

The second concern revolved around the responsibilities and behaviour behoved of the Board of Trustees by the aforementioned legislation as regards the extension. The *Imperial War Museum Act* 1920 itself was (and remains) a straightforward Act of Parliament. It continues to contribute towards the statutory framework constituting the museum, alongside other legislation such as the *Imperial War Museum Act* 1955 and the *Museums and Galleries Act* 1992[36] During the Second World War, the act comprised six sections. The first three were the most significant: section one defined the governing body of the Imperial War Museum—the Board of Trustees—and laid down the terms for its governance; section two set out the scope of the influence and power of the Board of Trustees, including vesting the Board with 'general management and control', thus rendering each Trustee a legal guardian; and section three established the core policies through which collection material could be gifted, acquired, or transferred to the museum. Sections four, five and six detailed staffing and expenses arrangements, the museum's non-charitable status and guidance on the legislation's citation and interpretation respectively. Although generally uninhibitive, one clause in section two did bestow officials from HM Treasury the power to prevent the purchase of potential exhibit pieces. This was clause 2d, stating that 'subject to the consent of the Treasury, [the Trustees may] apply any money received by them [...] in the purchase of any object which in the opinion of the Board it is desirable to acquire for the Museum'. In effect, the Treasury *could* veto any prospective acquisition made using government funding.

The significance of this clause cannot be underestimated. At many non-profit organisations such as museums, board members are often not just responsible for their institution's conduct, but also liable when that conduct causes negative ramifications. Charges against them can therefore arise from various contexts of perceived negligence. Douglas A. Johnston offers several examples from the museum sphere: 'Normally, board members are liable personally when they fail to fulfil one of the duties of all board members: (1) the duty of diligence; (2) the duty of obedience; and (3) the duty of loyalty'.[37] This, in turn, means trustees are subject to multiple dimensions of personal and collective accountability, with ramifications potentially befalling them if anything avoidable harms the organisation or contravenes its established governing framework.[38] Accordingly, Johnston recommends that: 'To avoid liability for actions beyond the scope of their duties, board members should conduct their duties strictly within their governing rules, whether in statute, charter or bylaws'.[39] The inference of this recommendation is that trustees may adopt, deliberately or otherwise, a conservative philosophy when dispensing their duties.

The above analysis indicates that the spectre of a separate if connected legal crisis had visited the Trustees, where the requirements of the Imperial War Museum contradicted its prevailing constitution.[40] This haunting was aggravated by concern for their situation should the legal crisis arise and be poorly managed. After all, as the institution's legal guardians, they, not Bradley, held ultimate responsibility for the Imperial War Museum. In doing so, it was also they, and not Bradley, who would be liable should the activity and any fallout thereof be met with disapproval from key governmental stakeholders or the process be undertaken unlawfully or against existing policies. The Trustees therefore felt the need for a higher political authority (the highest authority) to decide over the issue.

Following this outcome, the next step for Bradley should have been to raise the question about extending the subject remit of the Imperial War Museum with the Treasury. Instead, such was the caution which the Board of Trustees had immediately treated his proposal, the Chair of the Board, Sir Bertram Cubitt, instructed that its legality be considered first by the Treasury Solicitor's Office[41]—a subordinate Treasury department advising and acting for smaller government bodies, such as the national museums, which did not have their own legal team.[42] It should be noted here that the Chair's former connection to the War Office may also have influenced this instruction. From 1896 until 1926, Cubitt had held various senior civil service positions at the War Office, his last as Assistant Secretary.[43] Now retired, he still maintained contact with other senior War Office officials. This apparently further exposed Cubitt to concerns from the War Office about the *Imperial War Museum Act* 1920 and its limitations on the museum's subject remit, exposure which he raised with Bradley in his letter instructing the involvement of the Treasury Solicitor's Office.[44]

Bradley did as instructed and sent Cubitt's letter to the Treasury Solicitor's Office on 27 July 1939 along with a printed copy of the *Imperial War Museum Act* 1920. The lawyer who examined the case was

J. C. P. Kinsman. Unsurprisingly, his analysis found that 'there does not appear to be anything in the act to prevent the Trustees from accepting the custody of future relics of a future war'.[45] Highlighting, perhaps, the unnecessarily cautious behaviour exhibited by the Trustees, he appended his written judgment to Cubitt with the following pithy remark: 'I am at a loss to understand how the question ever arose'.[46]

This verdict stridently debunked key assumptions about the *raison d'être* and rationale of the Imperial War Museum. Despite contrary fears, nothing in the *Imperial War Museum Act* 1920 prescribed what the museum should or should not have collected, exhibited or interpreted. Reviews of contemporaneous Hansard proceedings, legislation and museum records carried out towards this study also indicate that until 1955, the act was its sole constitutional organ. Yet clearly, the Imperial War Museum had lost sight over this context. Through doing so, it became trapped in a paradox, simultaneously inhabiting two interrelated yet contrasting crises. At an important point during the resolution of the crisis of irrelevancy, the museum became distracted with the resolution of the apparent legal crisis, the actions mitigating the latter taking precedent over the actions mitigating the former. But the premise on which the sensed legal contradiction was founded had no substance whatsoever; there existed no binding apparatus that restricted collecting and display work to the First World War only. Alan Knight helpfully styles such situations as 'fictitious or contrived or, at the very least, hyperbolic [...]: those in which widespread fears are out of proportion to any objective threat'.[47] The reality of this particular situation here does not however delegitimise or underplay the potential crisis consciousness felt by the individuals who experienced it. After all, genuine concerns were afoot. While no documentation dictated the precise mission of the Imperial War Museum, other longstanding institutional markers explicated in Chap. 3 may have raised an impression of the rigidity of the subject remit. One was the institution's *raison d'être*, manifested by the prevailing collection amassed over its first 19 years, which had been dogmatically compiled from material pertaining to the First World War. Another was its formative, deep-rooted rationale, widely promoted since opening as a museum on the 'war to end all war'. And the third was its public profile generally, developed and solidified since 1920 as an institution concerning the affairs of 1914–1918. All the same, these points cannot take away from the fact that, in 1939, the Imperial War Museum had a large degree of autonomy, with any limitations it observed being entirely self-imposed.

The Imperial War Museum risked undermining its crisis management strategy through falling into this paradox. Unfounded concerns over lawful activity towards mitigating and extinguishing the threat of post-war cultural irrelevancy gave rise to conceptions and behaviours which stymied and ultimately waylaid the museum's crucial transformation work. Delays such as this can have serious detrimental effects for crisis management, especially in preparatory stages. They occupy time which might otherwise have been used to prevent, prepare for or reduce a crisis. When addressing sudden-onset situations, crisis discourse often emphasises the phase metaphorically known as the 'golden hour', which transpires immediately following crisis signal detection. This is a short time span of no specific arbitrary length wherein time-sensitive actions, necessary in resolving the crisis, must be performed.[48] Equivalent periods can also apply to slower developing crises or crises with longer lead-times, particularly those requiring extensive mitigation work such as the scenarios discussed during this study. Indeed, the Imperial War Museum squandered the golden hour in the resolution of the crisis of irrelevancy by becoming overly fixated on the legality of extending its subject remit, rather than with actually seeking the necessary permissions. Had the converse played out, collecting might have been possible from the moment the Second World War started. As events transpired however, it was not until well after hostilities began that the museum was sanctioned to begin this work.

Nineteen days after the declaration of war against Germany, and 85 days on from Kinsman's judgment over the *Imperial War Museum Act* 1920, Bradley was, finally, authorised by the Board of Trustees to approach the Treasury. The reason for the lull over July, August and September is unclear. In one account outlining the process by which the eventual permission had been secured, Bradley stated that: 'As soon as the evacuation of the most valuable pictures and photographs from the Museum was completed, I obtained the Chairman's permission to seek the decision which the Trustees asked for at their last meeting'.[49] But this statement masks a more complex picture. Contrary evidence shows that Bradley was eager to approach the Treasury sooner, suggesting he was aware of the problem which delays in this critical period might entail. On 7 July 1939, just over one week following Kinsman's judgment, the War Office informed him it too had concluded 'the Imperial War Museum is the proper body to deal with the preservation of the records and relics of any future war'.[50] Bradley responded by letter three days later, stating: 'I think personally it is very desirable to take steps to get our organization ready so that it can start to function at the outset'. He then closed with the following question: 'When, and how, can we start?'[51] No other evidence has been discovered wherein he expressed a desire to wait until after the evacuation of the Imperial War Museum. Whatever the reason for it, prolonged inactivity on this matter risked further diminishing irreparably the museum's contemporary collecting programme. Ultimately, the whole episode serves to highlight the significance of the mind in the creation of crisis, when even the approach to resolving a crisis is constrained by foundational assumptions raised out of the conceptualisation of another crisis.[52]

The Treasury received Bradley's formal request to extend the subject remit of the Imperial War Museum in a letter dated 23 September 1939. Ensuing memorandums between Treasury officials suggest no great exciting disagreement took place over the matter. For all the concern expressed by the Board of Trustees, it would appear that there was general consensus at the department about the benefit the extension would offer both the public and the Imperial War Museum. For example, one Treasury official reflected that: 'I think it must be admitted from all points of view that having set up the record of the Great War [...] it would be most desirable to continue with a similar record of the present war'.[53] Another mused how 'it is probably right to assume that future generations will be at least as interested in records of the present war as of the last war'.[54] And a third even predicted the feared outcome to this deep-revolutionary crisis were the requested permission not granted: 'Future generations will probably be as interested in this war as in the last war and would not attend in large numbers a Museum whose exhibits were limited to the last war'.[55] Consequently, on 19 October 1939, a document conveying permission was issued to the Imperial War Museum by the Lords Commissioners of the Treasury, a body which included the Prime Minister, Neville Chamberlain, as First Lord of the Treasury—the highest political authority possible.[56] Bradley laid this permission before the Trustees at the board meeting of 7 December 1939. Satisfied by its authority, the Board duly authorised him to begin organising the new collection.[57] In doing so, the first part of the preventative strategy towards addressing this crisis—seeking permission for the reinvention—had been concluded. The second part, executing the reinvention, had begun, and is explored over the next two chapters.

Challenging the Established Narrative on the Director-General

This chapter challenges an established narrative about Leslie Bradley built up over the past 25 years. While more recent research makes welcome interventions in the prevailing discourse surrounding him at the Imperial War Museum, much which remains understood about the man, his origins and directorship, is predicated on less well informed or self-serving stories occupying the realms of myth and legend rather than accurate biography.[58]

One quick example of this issue concerns Bradley's service record. Referencing his 1968 *Times* obituary, historian David Reynolds writes in *The Long Shadow: The Great War and the Twentieth Century*—a monograph about the impact of the First World War on the twentieth century—that Bradley was 'wounded at Ypres'.[59] This is incorrect. While the obituary does indeed disclose the fact that Bradley was 'invalided from the Ypres Salient'[60] (without any qualification for why), his service record shows the evacuation actually took place because he developed 'chronic infective condition of the mucus membrane of the large intestine'[61]—in current medical parlance, inflammatory bowel disease, typically comprising either Crohn's disease or ulcerative colitis[62]—rendering him 'unfit for further military service'.[63] But such comments are the least problematic inaccuracies about Bradley. Arguably the most problematic and personally damaging comments, which have also gone on to influence the way the Imperial War Museum, indeed IWM today, has viewed its history, originate from his successor, Noble Frankland.[64] This example concerns Bradley's approach towards the museum's direction and place in society as the Second World War loomed.

In 1994, while reviewing Gaynor Kavanagh's monograph *Museums and the First World War: A Social History*, Frankland lambasted Bradley's legacy when addressing the sections of the study that concerned the Imperial War Museum:

> The point she misses is that when the original object of helping to sustain morale in the war expired with the advent of peace, the new idea which took its place was that this museum would record and remind of a catastrophe which surely could never happen again. When, within a shockingly short time, it did, the original Imperial War Museum seemed to lose its mission. One of its most devoted servants, my immediate predecessor as Director,

> Mr L. R. Bradley, was quietly planning an institutional demise, which would be slow and dignified and would approximately coincide with his own.[65]

Two years later, in his memoir *History at War: The Campaigns of an Historian*, he pressed home this attack by reiterating the above accusation and buttressing it with another:

> Mr Bradley continued to see the Imperial War Museum through the eyes of its original beholders and those who founded it. [...] The coming of the Second World War shattered this illusion [...]. In that context, it was not wholly illogical for Mr Bradley to believe that the best course was to arrange for a gradual and dignified decline towards extinction for the museum which history had turned on its head. It did not occur to him that there was an alternative, or if it did, it was not one which he wished to embrace. He gave the impression of hoping, as he approached the grave, that the museum, which he had served for so long, would do the same.[66]

Included in this latter negative appraisal were broad statements about Bradley's character, particularly his historical subject interests and personal agenda for the museum:

> His empathy was almost wholly with the First World War [...]. To him, the Second World War was a nuisance which deposited masses of material in the Museum, squeezing its already restricted space and disrupting such order as its exhibitions had earlier had.[67]

The above portrayal of Bradley is at odds with the preceding and proceeding picture created herein from documentary evidence produced by Bradley himself, or other individuals about or concerning Bradley, during the Second World War. The information gleaned from material discovered in archives consulted towards this study belies Frankland's pointed criticisms.

It may be the case that Bradley eventually did become disillusioned with the Imperial War Museum during his final years as Director-General. This possibility is not disputed. A note written in 1960 certainly shows that the then Chair of the Board of Trustees had started perceiving Bradley's performance as 'beginning to slow up'.[68] He had not long turned 67 by this point. What is disputed, however, are Frankland's comments about Bradley's tenure during the Second World War era. As Bradley himself demonstrates through his own words from the period, he had nurtured the fundamental requirements for changing the museum's *raison d'être* and rationale as introduced in Chap. 2: dissatisfaction with the *status quo*, recognition of an alternative desired state and even some awareness of the necessary practical steps to obtain that paradigm, the overall process of which must be worth more than the cost of the change.[69] Indeed, Bradley was dissatisfied with the existing paradigm. He could see an alternative path before the Imperial War Museum which deviated from its original course. Moreover, Bradley was eager to direct the institution down this path and perceived that to do otherwise would be to condemn the museum to permanent closure. Consequently, far from endangering the Imperial War Museum, at the onset of the Second World War, he began taking the necessary action which evaded this crisis before it could bite, and bite hard.

Conclusion

The Second World War was the catalyst for an institutional transformation of the Imperial War Museum. From 1917 until 1939, the Imperial War Museum had embodied a museum on the 'war to end all war', as the First World War was then understood. The coming Second World War however threatened a deep-revolutionary crisis of irrelevancy. In response, the Director General, Bradley, decided that the museum needed to undergo major, wholesale change, framed herein as an institutional reinvention. This was so it could maintain pace with the shifting societal values. The reinvention process involved the Imperial War Museum transforming from a museum on the 'war to end all war' to a museum on the two world wars, effectively creating a new museum out of the one which already existed. Bradley became the key driving force behind this reinvention process. Contrary to prevailing stories surrounding his views on the scope and function of the museum, he perceived a credible future which involved extending the subject remit and representing stories and displaying material from the new world war. Bradley therefore raised this matter with the relevant authorities and worked towards securing the necessary permission for the project.

This first stage in preventing post-war cultural irrelevancy at the Imperial War Museum emphasises the importance of legitimacy to successful deep-revolutionary crisis management, indeed to all organisational matters. Alongside their activities and performance discussed in Chap. 5, museums must also ensure that an overarching deep-revolutionary crisis management strategy is fundamentally legitimate. This means any strategy which has been conceived to resolve a deep-revolutionary crisis-conducive situation should be consistent with society's prevailing socio-political requirements and regulations.

The Director-General had an acute awareness of and respect for the legitimacy requirements of the Imperial War Museum when responding to the anticipated threat of post-war cultural irrelevancy, most apparent through his (obligatory) indulgence of the concerns of the Board of Trustees about the appropriateness of the reinvention. Yet he also had significant concern and frustration with the delays which the the Trustees instigated for the chosen strategy. As this chapter shows, delays, whatever their origin, can have consequences for museums undergoing crisis management. Besides risking the situation from developing beyond control, they may thwart essential activities from being conducted on time or within tight windows of opportunity. This can restrict capacity at critical points during the process. It may also prevent activities being carried out in their entirety when needed. Accordingly, any uncertainty over the legitimacy requirements of a museum should, where possible, be settled as part of crisis management planning ahead of the onset of any crisis. Doing so will limit the potential delays that the work undertaken towards researching this information might otherwise cause. At the Imperial War Museum for example, uncertainty caused by an information-deficit in turn caused certain actors to sense—without foundation—a legal crisis-conducive situation which they felt threatened its socio-political regulative legitimacy. Addressing this illusory situation unnecessarily cost the museum time it could ill afford.

After securing permission for the reinvention of the Imperial War Museum, the second stage in preventing the post-war cultural irrelevancy was to deliver on it. This undertaking involved modifying the collection through accessioning material acquired from the new war and securing storage space for that fresh material, including through deaccessioning material which would unlikely ever be displayed. Before the necessary work could be accomplished however, the museum needed to establish the precise nature of its proposed reinvention. Such work involved establishing what the new, updated subject remit would comprise, as well as ensuring that the remit did not overlap with collecting initiatives by any other relevant museums. This activity is framed herein as conceptual reinvention, and comprises the subject of the next chapter in the ongoing exploration of the response of the museum to the anticipated threat of post-war cultural irrelevancy.

Notes

1. Montgomery Van Wart, 'The First Step in the Reinvention Process: Assessment', *Public Administration Review* 55 (1995): 429.
2. Brian Milstein, 'Thinking Politically About Crisis: A Pragmatist Perspective', *European Journal of Political Theory*, 14 (2015): 145; Reinhart Koselleck, 'Crisis', trans. Michaela W. Richter, *Journal of the History of Ideas* 67 (2006): 172.
3. IWM, Museum Archive (MA), EN2/1/COB/051/1, anon. to Melchett, 12 September 1939.
4. The National Archives (TNA), T 162/742/3, Bradley to the Secretary, 22 September 1939, 2.
5. Christine M. Pearson and Ian I. Mitroff, 'From Crisis Prone to Crisis Prepared: A Framework for Crisis Management', *The Executive* 7 (1993): 52–53.
6. Richard Beckhard, 'Strategies for Large System Change', *Sloan Management Review* 16 (1975): 45.
7. Edward S. Devlin, *Crisis Management Planning and Execution* (Boca Raton, FL: Auerbach, 2007), 10.
8. Milstein, 'Thinking Politically', 145.
9. Koselleck, 'Crisis', 172.
10. Keith Shaw and Louise Maythorne, 'Managing for Local Resilience: Towards a Strategic Approach', *Public Policy and Administration* 28 (2011): 46.
11. Pearson and Mitroff, 'From Crisis Prone', 52.
12. Christer Pursiainen, *The Crisis Management Cycle* (Abingdon: Routledge, 2018), 9–16; Ibrahim Al Luhaidan and Daifallah M. Alrazeeni, 'The Role of Key Early Warning Indicators in Crisis Management: Analytical Review', *Advanced Emergency Medicine* 8 (2019): 9–16.
13. IWM, MA, EN1/1/MUS/034, Bradley to Beards, 7 June 1939.
14. Milstein, 'Thinking Politically', 151–152.
15. Arjen Boin, Paul 't Hart, Eric Stern and Bengt Sundelius, *The Politics of Crisis Management: Public Leadership Under Pressure*, 2nd ed (Cambridge: Cambridge University Press, 2017), 23.
16. Ibid., 23–32.
17. Ibid., 39–40.
18. Emily Roe and Paul R. Schulman, *High Reliability Management: Operating on the Edge* (Stanford, CA: Stanford University Press, 2008); Gary Klein, *Sources of Power: How People Make Decisions*, pbk edn (Cambridge, MA: MIT Press, 1999); Daniel Kahneman and Gary Klein, 'Conditions for Intuitive Expertise: A Failure to Disagree', *American Psychologist* 64 (2009): 515–526; Daniel Kahneman, *Thinking, Fast and Slow* (London:

Penguin, 2011); Gene I. Rochlin, ed., *Journal of Contingencies and Crisis Management* 4 (1996): 55–112; Andrew Hopkins, *Learning from High Reliability Organisations* (Sydney, NSW: CCH Australia, 2009); Karl E. Weick and Kathleen M. Sutcliffe, *Managing the Unexpected Resilient Performance in an Age of Uncertainty*, 3rd edn (Hoboken, NJ: John Wiley, 2015).

19. Paul Cornish, '"Sacred Relics": Objects in the Imperial War Museum, 1917–1939', in *Matters of Conflict: Material Culture, Memory and the First World War*, ed Nicholas J. Saunders (Abingdon: Routledge, 2004), 35.
20. Owain Rhys and Zelda Baveystock, 'Introduction', in *Collecting the Contemporary: A Handbook for Social History Museums*, ed Owain Rhys and Zelda Baveystock (Edinburgh: MuseumsEtc, 2014).
21. TNA, AIR 2/10188, Bradley to Secretary of State for Air, 11 September 1939, pp. 1–2.
22. TNA, HO 186/2097, Bradley to Lord Privy Seal, 12 September 1939, 1–2.
23. TNA, HO 186/2097, Bradley to Secretary of State for the Home Department and Minister of Home Security, 14 September 1939, 1–2.
24. TNA, AIR 2/10188, Bradley to Ross, 24 October 1939, 1.
25. Beckhard, 'Strategies', 45.
26. Pearson and Mitroff, 'From Crisis Prone', 53.
27. Alexandros Paraskevas, 'Mitroff's Five Stages of Crisis Management', in *Encyclopedia of Crisis Management*, ed K. Bradley Penuel, Matt Statler and Ryan Hagen (Thousand Oaks, CA: SAGE, 2013), 629–630.
28. Boin et al., *Politics of Crisis*, 49–50.
29. Robert R. Janes, *Museums and the Paradox of Change: A Case Study in Urgent Adaption*, 3rd edn (Abingdon: Routledge, 2013), 44–50; Piotr Bienkowski and Hilary McGowan, *Managing Change in Museums and Galleries: A Practical Guide* (Abingdon: Routledge, 2021), 123.
30. IWM, MA, EN1/1/COB/049/2, Board of Trustees, 20 June 1939, 2.
31. Ibid.
32. Ibid.
33. Ibid.
34. Monica A. Zimmerman and Gerald J. Zeitz, 'Beyond Survival: Achieving New Venture Growth by Building Legitimacy', *Academy of Management Review* 27 (2002): 418.
35. Ibid.
36. *Imperial War Museum Act*, 1955, c. 14; *Museums and Galleries Act*, 1992, c. 44; IWM, 'Annual Report and Accounts, 2022–2023' ([London?]: HMSO, 2023), 60, https://assets.publishing.service.gov.uk/government/uploads/system/uploads/attachment_data/file/1178402/Imperial_War_Museum_Annual_Report_and_Accounts_2022_to_2023.pdf (18 March 2024); see IWM, 'Annual Report and Accounts, 2020–2021'

([London?]: HMSO, 2021), 92, https://assets.publishing.service.gov.uk/media/61bcbf50d3bf7f054e76962c/CCS0921374640-001__Imperial_War_Museum_Web_Accessible__5_pdf (18 March 2024).

37. Douglas A. Johnston, 'Directors and Trustees: Personal Liability for Their Actions', *The International Journal of Museum Management and Curatorship* 7 (1988): 77.
38. G. Ellis Burcaw, *Introduction to Museum Work*, 3rd edn (Walnut Creek, CA: AltaMira, 1997), 205.
39. Johnston, 'Directors and Trustees', 78.
40. See Philip W. Deans and Darren Kelsey, 'National Narration and Organisational Crisis: A Discourse-Mythological Analysis of the Imperial War Museum During the Second World War', *CADAAD Journal* 15 (2024): 86–108.
41. IWM, MA, EN1/1/MUS/034, Secretary to Kinsman, 27 June 1939.
42. Norman Wilding and Philip Laundy, *An Encyclopaedia of Parliament*, 3rd edn (New York, N.Y.: Frederick A. Praeger, 1968), s.v. 'treasury'.
43. 'Sir Bertram Cubitt', *Times*, 24 September 1942.
44. IWM, MA, EN1/1/MUS/034, Cubitt to Bradley, 23 June 1939.
45. IWM, MA, EN1/1/MUS/034, Kinsman to Bradley, 29 June 1939.
46. IWM, MA, EN1/1/MUS/034, Cubitt to Bradley, 23 June 1939.
47. Alan Knight, '"Crisis" and the Great Depression in Latin America', in *Crises in Economic and Social History: A Comparative Perspective*, ed A. T. Brown, Andy Burn and Rob Doherty (Woodbridge: Boydell, 2015), 35.
48. Helio Fred Garcia, 'Effective Leadership Response to Crisis', *Strategy and Leadership* 34 (2006): 4.
49. IWM, MA, EN1/1/MUS/034, template for correspondence from Bradley to the Trustees.
50. IWM, MA, EN1/1/MUS/034, Redman to Bradley, 7 July 1939.
51. IWM, MA, EN1/1/MUS/034, Bradley to Redman, 10 July 1939, 1–2.
52. Deans and Kelsey, 'National Narration'.
53. TNA, T 162/742/3, memorandum, by an unidentifiable individual, 30 September 1939, 2.
54. TNA, T 162/742/3, memorandum, by an unidentifiable individual, 4 October 1939.
55. TNA, T 162/742/3, memorandum, by an unidentifiable individual, 9 October 1939.
56. IWM, MA, EN1/1/MUS/033, Douglas to The Duke of Gloucester, Lords of the Treasury, Trustees of the Imperial War Museum and Bradley, 19 October 1939.
57. IWM, MA, EN1/1/COB/049/4, Board of Trustees, 7 December 1939, 1–2.

58. Jim Aulich and John Hewitt, *Seduction or Instruction? First World War Posters in Britain and Europe* (Manchester: Manchester University Press, 2007); Alys Cundy, 'A Century of Reinvention: Display Policy and Practice at the Imperial War Museum, London, 1917-2017' (doctoral thesis, University of Bristol, 2015); Leanne Green, 'Advertising War: Pictorial Publicity, 1914–1918' (doctoral thesis, Manchester Metropolitan University, 2015).
59. David Reynolds, *The Long Shadow: The Great War and the Twentieth Century*, pbk edn (London: Simon and Schuster, 2014), 244; 'Mr L. R. Bradley', *Times*, 30 January 1968.
60. 'Mr L. R. Bradley', *Times*.
61. TNA, WO 339/54103, 'Proceedings of A Medical Board', 25 August 1916.
62. See Louise Langmead and Peter Irving, *Inflammatory Bowel Disease* (Oxford: Oxford University Press, 2008).
63. TNA, WO 339/54103, 'Proceedings of A Medical Board', 25 August 1916.
64. Terry Charman, '"A Museum of Man's Greatest Lunatic Folly": The Imperial War Museum and its Commemoration of the Great War, 1917–2008', in *A Part of History: Aspects of the British Experience of the Western Front* (London: Continuum, 2008), 99–106; James Taylor, 'Interpreting the Second World War', in *The Power of Objects: Museums and World War II*, ed Esben Kjeldbæk (Edinburgh: MuseumsEtc, 2009), 52–81.
65. Noble Frankland, review of *Museums and the First World War: A Social History*, by Gaynor Kavanagh, *Journal of the History of Collections* 7 (1995): 127.
66. Noble Frankland, *History at War: The Campaigns of an Historian* (London: Giles de la Mare, 1998), 164.
67. Ibid., 163–164.
68. TNA, T 218/207, 'Note for the Record', by Loughnane.
69. Beckhard, 'Strategies', 45.

CHAPTER 7

Planning the Reconception, 1939–1941

Introduction

The final two chapters herein advance the critical and historical narrative embarked on in Chap. 6 regarding the work by the Imperial War Museum to resolve a deep-revolutionary crisis of irrelevancy. Whereas Chap. 6 explored how the museum detected the crisis signals forewarning this problem, sensed the significance it had for the institution, and conceived the necessary resolution and mitigating measures, Chaps. 7 and 8 examine processes underlying the chosen strategy. Together, they mount an argument that museums confronting deep-revolutionary crisis should develop away from danger by undergoing institutional reinvention: creating something new from something which already exists, as discussed in Chap. 2. The current chapter contributes to this argument by examining the particular process which defined the reconception of the Imperial War Museum. This includes considering the external pressures and interventions alongside those originating from within. In doing so, the chapter contends that the reinvention of a museum first necessitates conceptualising precisely what the reinvented museum will look like. It also contends that this conception can arise from both internal and external forces.

The historical context of the current chapter is a rivalry between the Imperial War Museum and another national institution during those most precarious war years for the United Kingdom, 1939–1941. Before this

P. W. Deans, *Crisis, Reinvention and Resilience in Museums*,
https://doi.org/10.1007/978-3-031-67806-6_7

period until 1934, the Imperial War Museum enjoyed an effective monopoly on material pertaining to the First World War. Only two other institutions possessed a remit with overlap: the Science Museum and Royal United Service Institute Museum. The extent these other institutions actually overlapped was very limited. The Science Museum did not concern itself 'with war relics as such except when these relics happen to illustrate some outstanding technological advance or invention'.[1] And while the Royal United Service Institute Museum 'concentrate[d] as far as possible on [collecting] small trophies or mementoes of famous actions or events [...] [involving] His Majesty's Armed Forces (including Dominion, Indian and Colonial Forces)',[2] in reality, this took place, as claimed by the Imperial War Museum, 'on a comparatively small scale', with 'a cordial collaboration [existing] between [both institutions]'.[3] The many regimental museums founded before and after the First World War were also mostly interested in subjects and events concerning their regiment only.[4] By 1934 however, a new museum had come into existence which did significantly overlap the Imperial War Museum on matters pertaining to war at sea.[5] This was the National Maritime Museum, emerging from the national museum investment programme of the 1930s, concerned with the 'preservation and exhibition of objects connected with the Naval and Mercantile Marine Services, harbours and ports, voyages of exploration and discovery etc. and portraits etc. connected therewith'.[6] During the Second World War, the Imperial War Museum and the National Maritime Museum found themselves competing for Naval exhibits, requiring an understanding to be developed about how the former would represent the Second World War, which set it apart from the latter.

Understanding the reinvention of the Imperial War Museum during the Second World War era requires this concept to be articulated over the next two chapters as a two-stage change process, conceived from ideas by reinvention theorist Willie Pietersen.[7] The first stage involves theoretical work delineating what a changing museum's desired paradigm should look like. The second involves practical work realising this new paradigm through bringing the museum's various organisational elements into alignment with it. Hence, the current chapter draws on what has been conceived herein as 'conceptual reinvention': the process of establishing new *raison d'être* and rationale. Similarly to Chaps. 5 and 6, all this is considered within framing provided by Christine M. Pearson and Ian I. Mitroff's general five-stage framework for crisis management, specifically its second 'prevention' stage.[8] The current chapter also draws on the concept of the 'spectacle' in analysing the conceptual reinvention undergone by the Imperial War Museum. This is to interpret a descriptor historically applied against the museum during the conceptual reinvention process.

The current chapter addresses the above over three substantive sections. The first section introduces and explicates conceptual reinvention. It elucidates the concept and any relevant associated issues, ideas and activities. The second section then discusses a gradual shift in curatorial approach at the Imperial War Museum following the Second World War. It contrasts this post-Second World War approach with the one followed before the conflict and surmises why the change was permitted. And the third section explores the catalyst for that change, specifically the conceptual reinvention which the Imperial War Museum underwent in response to the crisis-conducive situation faced by the museum, arising from the perceived threat of post-war cultural irrelevancy. It sets out the conceptual reinvention process, investigates the outcome and considers an unsolicited intervention by the National Maritime Museum, which partly influenced the conceptual reinvention's course and outcome.

Conceptual Reinvention

Reinvention is the ultimate response to the threat—real or imagined—of an institution becoming irrelevant in the social system. It concerns organisational purpose as well as organisational structures and practices. A museum is no exception. Here, reinvention transcends physical elements to effect non-physical elements too, such as the museum's mission, values and vision: three notions which guide museum purpose, principles, practice and outlook. The mission drives a museum forward. It comprises an articulated understanding conveying and embodying the museum's *raison d'être*, rationale and aims and objectives. The mission is informed by a museum's values. Values represent the issues and ideas which inspire the museum to action. And the vision is the desired outcome of museum work: a forward projection, taking into account its mission, values and the prevailing social system.[9] This understanding has implications for what museum reinvention means and entails. As Gail Anderson puts it: 'Reinventing the museum is not just adding a program, reinstalling a gallery, or increasing financial reserves—it is a systemic shift in attitude, purpose, alignment, and execution'.[10] Accordingly, the reinvention of a museum concerns more than just merely retooling what the museum does; it involves redefining what the museum is and is for.

This conceptual reinvention of museums typically begins at the relevant institution as a response to dissatisfaction with the *status quo* through, for example, concerns about encroaching irrelevancy.[11] It necessitates critical self-examination: asking and answering hard questions about what the museum is, is for and does. In doings so, the changing museum can establish what the desired state actually is.[12] This process has become known as a situational analysis, embodying a critical reflection towards analysing all the internal and external factors to a museum.[13] Pietersen thematises the various issues, ideas and activities comprising this opening stage of the reinvention process as 'learning'.[14] In the museum context, Gary Edson and David Dean propose various introspective questions to support this work.[15] They include questions geared towards understanding issues including the unique role of the relevant museum, who its constituents are, what they and the wider community needs, and the relationship between the museum and them, *etcetera*. Through making assessments and revealing the disparities between their organisational contexts and the societal values they must negotiate and factor, museums can determine the destination for their reinvention.

Following situational analysis, conceptual reinvention requires a period wherein the precise nature of the desired reinvention process is clarified. Pietersen thematises the issues, ideas and activities comprising this work as 'focusing'.[16] Without an understanding of what the precise aims and objectives comprising some reinvention process actually is, reinvention cannot take place safely or effectively. It risks occurring without clear direction and, in doing so, risks becoming a drain on resources without anything useful or productive to show for the expenditure.[17] As per all organisations, museum user expectations and requirements lie at the heart of museum reinvention. Drawing on the information derived from a situational analysis, museums must, where practicable, implement an offer that delivers what users seek from museums.[18] The focussing of reinvention may otherwise end up delivering something which is neither wanted nor needed. Visitor research has an important part to play in that process. Depending on the methodology deployed, it can help ascertain what motivates attendance at museums when other recreational activities vie for attention. This possibility makes visitor research an essential task in the process of focussing reinvention.[19]

Another important consideration to make when focussing reinvention is the museum's service and competition. All organisations, whatever their *raison d'être* and rationale, must fill some specific societal need. This requirement is met by offering what Pietersen calls a 'winning proposition'.[20] In the commercial context, winning propositions comprise some public offer that retains or enhances an organisation's unique selling point, enabling them to stand out from their competitors—at the competitor's expense, if necessary. Similar thinking can be applied against museums, which also need emphasis.[21] The difference with the museum sector, however, is that, as public service organisations with limited resources, museums should (ideally) work together rather than against each other. Consequently, unlike commercial organisations, they should consider the interests of other existing or potentially comparative museums when undertaking strategic planning, as encouraged by articles 3.9 and 3.10 of the International Council of Museums' current *Code of Ethics for Museums*.[22] To do otherwise potentialises what is known in museum discourse as 'duplication of effort'. This phenomenon occurs when two or more opposing museums hold the same mission, values and vision or pursue similar aims and objectives with limited resources.[23] As such, while some winning proposition is important for museum reinvention, a museum reinvention which does not take into account the *raison d'être* and rationale of other museums risks being detrimental to prevailing museum services and the local, regional or national museum ecology.[24]

A 'Spectacular' Museum

Before considering the conceptual reinvention of the Imperial War Museum during the Second World War and exploring the extent that the National Maritime Museum influenced this outcome, it is necessary to obtain an understanding of the development trend of the institution's post-war curatorial framework and how that framework differed from the one followed before the Second World War. Otherwise, the findings presented in the current chapter risk losing significance. Following the end of the Second World War, the Imperial War Museum—or IWM, since 2011—increasingly turned away from a commemorative framework towards profiling as well as displaying the technology brought to battle by fighting militaries over the twentieth and, later, the twenty-first centuries, particularly large and imposing examples such as vehicles, aircraft, boats and other equipment. The 2009 poster campaign which depicted silhouettes representing various iconic technological objects from its collection illustrates this shift. IWM curator Paul Cornish explains that particular campaign responded to visitor surveys indicating they 'were the most attractive element of the IWM's collection to children'.[25] Over the past seven decades, the 'wow' factor, and the emotional responses it evokes, have become key pillars of the museum's visiting experience.

Such public interest in *matériel* is clearly helpful for any museum that displays the tools and effects of war and armed conflict, particularly in cultural contexts where killing machines are regarded as the peak of technological development and the ultimate expression of state and human power.[26] It adds a certain cachet, which can support and enhance corporate and particularly curatorial activities. The apparent allurement of *matériel* to prospective visitors at institutions like the Imperial War Museum/IWM stems from what Theodor W. Adorno observes as people's predisposition and attunement with technology.[27] This receptiveness has occurred through many societies the world over achieving increasingly technologically advanced conditions. Adorno goes on to posit that in doing so, people have become fixated with technological impact rather than effect: 'People are inclined to take technology to be the thing itself, as an end in itself'.[28] This implies a degree of fetishisation amongst developed societies over technology and the capability and power it affords the user, which museums can readily factor into their interpretation and promotion.

But if war museums exhibit technological objects to satisfy this interest, any corporate justifications for doing so (should) exist secondly to the intellectual rationale behind why they are collected, displayed and interpreted. When viewed through a long temporal lens, human conflict since the First World War has been uniquely destructive, killing not just people, but also increasingly destroying material culture, architecture and even whole landscapes.[29] This has come about, explains Cornish, through scientific advancement and technological development: 'The industrialisation of war was the salient feature of both world wars—it was technological and industrial progress that gave them their shockingly lethal character and permitted them to be waged on such an unparalleled scale'.[30] A more meaningful, ethical reason for displaying military technology in war exhibitions, therefore, is their ability to springboard issues and ideas about the furtherance of human ability and all that entails during war and armed conflict: the lethal extension of human dexterity.[31] As Cornish argues: 'One cannot attempt to represent modern conflict without showing the weapon-systems used to wage it'.[32] Considering this, it is behoved of institutions like the Imperial War Museum/IWM to conserve and exhibit even the ghastliest weaponry.

Through increasingly profiling the technology of war in its displays and interpretation following the Second World War, the Imperial War Museum became increasingly associated with what has been historically described as the 'spectacular': spectacular objects, spectacular displays and spectacular experiences. Documentary evidence supports this. Drawing on archival records, reports and publications, the current chapter aims to show a clear direction of travel for the museum, promoted by the Standing Commission on Museums and Galleries.

Beginning with its meaning, 'spectacular' derives from the word 'spectacle', which Charity M. Counts defines as 'an event or experience that one simply cannot help but look at, whether in wonderment, curiosity, shock or awe'.[33] Various cultural theorists including Guy Debord, Anne McClintock, Tony Bennett and Nick Prior have engaged with this concept during discussion on the appropriation of imagery to convey state, colonial, technical and commercial power respectively.[34] In an exhibition context, the spectacular could concern pictures, objects big and small and intangible constructs such as ideas, concepts and cultures. And while historically it may have conjured feelings of sensationalism and jingoism amongst exhibition attendees, IWM shows this is not the only rationale for doing so.[35]

Harnessing the spectacular qualities of its most impactful objects, whether technologically related or not, is an interpretative approach now matured by IWM and deployed sensitively to positive effect. As Kasia Tomasiewicz notes, this has included ensuring 'the active effects of war on objects are not obscured but valued components of their materiality'.[36] By doing so, such exhibits become what Francesca Monti and Suzanne Keene call 'star' and 'gateway objects'.[37] That is, they comprise objects which, through their spectacularness, create opportunities to learn about and explore related contexts and neighbouring exhibits. One recent example of this can be seen with the large, iconic objects exhibited in the 'Witness to War' atrium display at the museum's London headquarters following the site's First World War centenary regeneration. Other examples include objects at the Duxford site, particularly in Air Space, the American Air Museum and the Land Warfare Hall, and at the Salford site. All this helps uphold what James Scott calls a realistic interpretation of *matériel* as opposed to one that celebrates and venerates.[38] Fundamentally, it is an approach which contrasts with the one pursued before the Second World War. As discussed in Chap. 3, prior, the Imperial War Museum followed a commemorative framework. Exhibits, including large objects, were presented as possessing connections with the country's national and imperial war dead, giving them memorial-like qualities. Any other curatorial decisions or remarks applied against them were incidental to this framework.

How the Imperial War Museum embarked on the journey towards its current interpretive paradigm has been explained in prevailing scholarship as being a product of the institution's post-Second World War development. The interpretation is seen thenceforth to focus increasingly on technological abilities and effects until it became the dominant theme. Research by Alys Cundy shows that, through doing so, the Imperial War Museum slowly cast off the initial commemorative framework and adopted a technological framework instead.[39] One early marker of this, identified by Cundy, concerns the interpretation regarding an ordinary tile, '"blistered" by [the atomic bomb's] heat flash at Nagasaki', that the museum acquired in 1950. The interpretation for the tile focussed on why the blast had impacted the tile so, but entirely omitted to consider any effects which the blast would have caused people.[40] Another, later marker, identified by Cornish, concerns the development of the first (smaller) atrium at the museum's London headquarters in 1989, built for exhibiting large objects near the entrance where the army and navy galleries had previously existed. Originally, the interpretation used there was 'strictly limited to the technology on show'.[41]

A major reason for the turn of the Imperial War Museum towards a technological framework after the Second World War has been given in its first post-war guidebook as deriving from visitor demand to see the 'many new equipments evolved during 1939-1945, of which they had heard but had had few opportunities of seeing'—besides the general if limited post-war interest in the Second World War following the conflict.[42] Similar views are also documented around this time by the new Director of the Science Museum, Herman Shaw, who wrote for the December 1946 edition of *Museums Journal*:

> There is no doubt whatever, that during the last few years, probably as a result of war-time developments in science and engineering, the public has become increasingly conscious of the impact of science on their everyday life, and are clamouring for authoritative information regarding all branches of science and scientific research.[43]

This coheres with what Adorno notes as society's increasing interest with technology.[44] But such sources and ideas delineate only half the picture. Granted, they do help explain *why* the Imperial War Museum broke from its commemorative framework following the Second World War: perceptions of bottom-up, public demand. Yet they do not help explain *how* it could have realised this shift after following the former interpretative paradigm since 1920. Change of this nature requires significant institutional reframing. It necessitates a strategy which would require legitimisation from all the museum's stakeholders[45]—defined here as anyone with interest in an organisation, project or decision.[46] Through next considering the conceptual reinvention of the Imperial War Museum during the Second World War, the current chapter establishes the process by which this legitimisation came about.

Resolving a Crisis of Irrelevancy at the Imperial War Museum (II): Establishing the Response

Jostling for Exhibits with the National Maritime Museum

During the Second World War, the Imperial War Museum and the National Maritime Museum found themselves competing for naval exhibits. This rivalry was unintentionally initiated on 10 September 1939, when the Director-General of the Imperial War Museum, Leslie Bradley, wrote to the Admiralty seeking support for the institution's extension. His letter stated: 'all members of the staff will, if given the opportunity, spare no effort to ensure that the work and efforts of the Royal Navy are adequately recorded'.[47] The Royal United Service Institute Museum also made similar overtones the following month.[48] These requests seemingly aroused the attention of the National Maritime Museum and caught it completely by surprise. The Trustees there had not previously felt the need to stake a claim. Rather, having understood that their institution 'was established by Act of Parliament in order to portray the deeds of the Navy and the Merchant and Fishing Fleets throughout [...] history', they had assumed the National Maritime Museum would be given first refusal over objects.[49]

In response, the Director of the National Maritime Museum, Sir Geoffrey Callender, wrote an urgent letter to the Admiralty seeking reassurance. He stated that his Trustees:

> view[ed] with dismay any counter proposals which denied to them the power of linking up the history of the Royal Navy in our own time with the history of the Royal Navy in the days of Drake, Blake and Nelson.[50]

Callender makes a very convincing argument in this letter for why naval material should be deposited at the National Maritime Museum ahead of other institutions. As he explained:

> although the Act [...] inaugurating the Imperial War Museum preceded the National Maritime Museum Act by fourteen years, the National Maritime Museum Act Section 3 sub-section (a) [stating 'which at the time of the constitution of the Board form part of the Naval Museum of the Royal Naval College'] recognised the Royal Naval Museum [...] which traces its origin back [...] prior to 1832. In transferring [...] the contents of the Royal Naval Museum, the Act of 1934 demonstrated [...] the lines on which the National Maritime Museum should be conducted.[51]

Interestingly, and despite making this convincing argument, it appears that Callender felt rather insecure about the position of the National Maritime Museum, particularly as regards the Imperial War Museum. This is surmised because a month later, he followed up with another letter seeking to aggressively undermine the claim made by the Imperial War Museum. Indeed, Callender asserted (incorrectly, as demonstrated in Chap. 6) that any extension of the subject remit of the Imperial War Museum would require being 'brought before Parliament' and that, because 'the War Museum is "Imperial" in character, [...] any amendment to the [*Imperial War Museum*] *Act* of 1920 will [also need to] be brought to the notice of other Museums, not only in this country, but in the Dominions and Colonies' (italics added).[52] He did however buttress these assertions by asserting correctly that 'The Imperial War Museum is already so congested that an extension of its range would necessitate a considerable reduction of the 1914–1918 exhibits, unless an ambitious building scheme were found practicable'[53]—an issue for the institution which is considered during Chap. 8.

The position and tone that Callender took on this matter contrasts strikingly with Bradley's. At no point on record does Bradley seem to have dismissed outright the claim of the National Maritime Museum. Rather, he simply cautioned that 'it would overbalance that Museum if it attempted to deal with the Naval Relics of this war on the same scale as the Imperial War Museum'.[54] This is because, before the Second World War, the

National Maritime Museum had a subject remit which covered over one hundred years of British maritime history. The Imperial War Museum, by contrast, focussed on a mere four. Indeed, Callender's assertive conduct over the issue goes some way to explaining, if not justifying, an extremely hard character reference of him made by the then Director of the Science Museum, Ernest Mackintosh—himself a Trustee of the Imperial War Museum. Writing to Bradley around the same time, Mackintosh described Callender as 'a perfect Nazi—shifty, grasping, opportunist, and unscrupulous [...]. He alters the scope and prerogative of the [...] [National Maritime Museum] to suit his whim or covetousness'.[55] To be absolutely clear, Mackintosh's use of the term 'Nazi' in this context was unlikely an accusation regarding Callender's political ideology. As Nigel Rees explains, during the Second World War, Nazi became *the* hate word in the United Kingdom: 'as though it related to nasty'.[56]

Through these opposing requests, the Treasury, in concert with the Admiralty, War Office and Air Ministry, decided that the most appropriate action for dealing with the situation was to consider the matter holistically. That is, decisions on their allocation would be linked up with considerations regarding ground- and aerial-related material as well. Such was the complexity of and specialist knowledge required for the task, the Treasury passed this undertaking over to the Standing Commission on Museums and Galleries, requesting the Standing Commission 'consider and report upon the most appropriate allocation of relics of the present war between the institutions concerned'.[57] These included the Imperial War Museum, National Maritime Museum, Science Museum and the Royal United Service Institute Museum. The potential acquisitions which would be earmarked by the service departments for collection after the war could then be distributed centrally as per the report's recommendations.[58] This idea conciliated Callender, who on receiving the news responded by stating: 'I am sure that all members of the Board will welcome the decision to transfer responsibility in this matter'.[59] The Standing Commission therefore created a subcommittee to deliberate over the question of the allocation of material amongst these institutions.[60]

The Standing Commission on Museums and Galleries Material Allocation Subcommittee

Throughout 1940–1941, evidence was submitted to the Standing Commission by these four institutions, relevant government departments and the War Artists Advisory Committee—the latter comprising a wartime body tasked with commissioning and disseminating war art amongst the country's national and local institutions.[61] Submissions included letters, memoranda and oral testimony, establishing the war-related collecting requirements, or views on them, of the various aforementioned institutions. The evidence was then deliberated over by what shall be termed here as the Standing Commission's 'material allocation subcommittee', which seemingly never received an official name.[62] This process resulted in a report being delivered on 14 August 1941 offering recommendations for the distribution of army, navy and air force related material amongst the four institutions.[63]

Claims from the Science Museum and the Royal United Service Institute Museum were quickly and easily dealt with by the allocation subcommittee. The Science Museum possessed no interest in conflict *per se*. It was only concerned with acquiring objects of significant scientific interest.[64] Accordingly, the report accepted the museum's claim for material such as 'aircraft models and components, engines, scientific instruments, wireless apparatus, vehicles, etc. etc. [...] which it would have been interested [in] had the same scientific and technical developments occurred in the absence of war'.[65] Moreover, the Royal United Service Institute Museum, which was interested in obtaining exhibits derived from the war involving the United Kingdom and the British Empire, had somewhat restricted scope and space. Hence, it only sought 'small trophies or mementoes of famous actions [...] and [significant] personal relics'.[66] The allocation subcommittee therefore similarly concluded that: 'We see no objection to the Institution including [...] [these] in its collections on the same lines as those of previous wars'.[67] But the Imperial War Museum and National Maritime Museum, by contrast, proved wholly more difficult matters. Questions over the disposal of the naval material between these museums presented awkward issues that occupied the subcommittee's time. Furthermore, the conclusions drawn had broad consequences for the Imperial War Museum—the focussing of its reinvention—which catalysed changes to the institution's identity and curatorial policies as discussed above.

The Imperial War Museum submitted three pieces of evidence to the allocation subcommittee when making its case for receiving material from the service departments. Through these submissions, the museum demonstrates what Timothy Ambrose and Crispin Paine call a 'forward plan'.[68] Drawing on findings from situational analysis, these provide museums with the imagination and framework to improve organisational effectiveness and resolve any problems which might exist.[69] The first piece of evidence was a letter from the then Chairman of the Board of Trustees, Sir Bertram Cubitt. This summarised the Trustees' position on the scope maintained by the Imperial War Museum and the material it felt should be deposited there.[70] The second was an extended memorandum by Bradley. This explicated the issues and ideas conveyed in Cubitt's aforementioned letter and established the museum's aims and objectives and its expectations.[71] And the third was oral testimony via an interview which the Standing Commission conducted with Bradley. This sought to clarify or further develop on points raised in the two previous sources.[72] The evidence was also appended with a list exemplifying the naval material which the Imperial War Museum sought, plus copies of letters from the Treasury authorising the institution's extension and requesting that material be set aside for it by the various government departments.[73] This forward plan indicates a reflective learning process had been undertaken at the museum towards understanding its *raison d'être* and rationale.[74] At the same time, the plan offers an insight into the museum's vision for, and hence focus of, the then proposed reinvention.[75] In doing so, the plan helped develop comprehension about the desired state which the museum sought.[76]

Notwithstanding its newly sanctioned subject remit, the argument which the Imperial War Museum put forward drew on the understanding that the institution was one of popular appeal, specifically in two contexts. One was representational, the other demographic. The first context derived from a view that the institution had been established to commemorate the efforts made by every participating person from the United Kingdom and British Empire during the previous world war.[77] As Cubitt's letter informed the subcommittee, the Imperial War Museum existed 'to form a memorial and record of the effort and sacrifices of the empire in all aspects of the war'.[78] In line with this, Bradley's memorandum contended:

> [it] must continue to have [represented] within its scope the work of the Royal Navy, Merchant Navy, Army, and Air Force, of the men and women of the Civil Defence, the munition workers, the women's services, and the Dominions, and to deal with the effects of the war on the ordinary civilian, and all its endless repercussions.[79]

The second demographic context derived from the aforementioned representational context. By being ostensibly an institution that thitherto had represented the work of everybody who contributed to and participated in the First World War, the Imperial War Museum aimed its representation, so the memorandum claimed, at 'the "popular" visitor, rather than the technical, as distinguished from the general, student'.[80] Put simply, the museum's content was focussed on catering for the everyday person, rather than the specialist. Jennifer Wellington demonstrates this to have been a longstanding trait of the Imperial War Museum, aligned with its deliberate, founding popular appeal discussed in Chap. 3.[81] When the general public began visiting after opening at the Crystal Palace, curators noted how many people expressed 'surprise' that mundane and everyday objects would be considered 'worthy of preservation'. As Cundy explains, this curatorial decision was deliberate: 'The IWM's founders sought to differentiate themselves from [...] more restricted collections by presenting the museum as a place where those who had experienced the war in any capacity could easily visit and see themselves represented'.[82] In developing the point during his oral testimony towards a discussion on whether the Imperial War Museum acquired scientifically related collection items, Bradley concurred with an assertion put forward by the Standing Commission member Sir Albert Steward that the Imperial War Museum was:

> not interested in scientific instruments as such, but [...] [was] interested in a scientific instrument which, through its wide application and importance in the war, became familiar by name to the public, or in any scientific instrument which became famous through its connection with a particular person or a particular event.[83]

The technical qualities of individual exhibits and the displays overall had previously been emphasised in publicity literature to help popularise the museum to the public.[84]

Through these submissions, the Imperial War Museum appealed to the allocation subcommittee for exhibits prevalent in societal consciousness and representative of both the civilian and military war effort. They show the desired reinvention was being geared towards addressing the Second World War through a *raison d'être* and rationale which broadly cohered with the institution's founding ideals: mass communication, commemoration and representation—although now minus the conceit that the previous world war had been the 'war to end all war'. The left-hand column of Table 7.1 (below), listing the naval material that the Imperial War Museum sought, illustrates this. Desired items included equipment or models

Table 7.1 Two lists of naval material recommended for the Imperial War Museum

'Short List of Naval Material Required for the Imperial War Museum Based on the Contents of the Present Naval Galleries and the Reference Departments'	*'Categories Suitable for the Imperial War Museum'*
Photographs of ships of the period. Classified list of H.M. Ships at outbreak and end of hostilities: statistics of enlistments, casualties, awards, etc. Builders models of famous ships, including Fleet Auxiliaries. Relics and actions of famous ships. Models illustrating actions and methods of attack and defence. Historic ordinance and special weapons and means of defence evolved during the war. Naval shells, fuses, and cartridges. Naval uniforms, equipment, and badges of rank and proficiency (including Royal Marines and Fleet Air Arm). Ships' badges. Mine warfare, including British and German mines, minelaying and minesweeping. Submarine warfare, including depth charge, asdig, hydrophone, torpedo, and defensive arming of merchant ships. Convoy. Anti-aircraft defence. Merchant Navy; photographs, models, and relics: defensive armament and equipment; statistics. Navy Medical Service: models, equipment, methods. Fleet Air Arm; models of naval aircraft, bombs, equipment, armament. Dominion Fleets; relics of actions, photographs, statistics, uniforms. Allied Fleets; photographs, uniforms, relics of actions. Enemy Fleets; captured trophies, photographs. Pictures, drawings, portraits, busts. Commemorative medallions. Charts. Photographs and films. Ship magazines and journals; official histories; published books. Admiralty publications, e.g. orders and regulations; strengths of fleets, returns of losses, navy lists, handbooks relating to weapons and ammunition.	Artillery; except in the form of models, designs, and measured drawings. Lethal weapons other than cannon; except in the form of models, designs and measured drawings. Torpedoes, Contact Mines, Magnetic Mines. Depth-Charges Actual Guns preserved as trophies of personal heroism, e.g. guns served by Jack Cornwell, V.C. at Jutland; or last in action in H.M.S. Hardy at Narvik. 18" shells and other projectiles. Full-sized paravanes, and range finders. Armour Plating for H.M. Ships. 1. Actual Samples 2. Specimens subjected to enemy's fire. Fragments of Ships, preserved as trophies of personal valour, etc. Relics of memorable ships, e.g. Lusitania. Models of famous ships of R.N. and M.M. other than such as are constructed exactly to scale. Care of Sick and Wounded. 1. Hospital Ships and Base Hospitals: models, etc. 2. Uniforms and Nursing Service. 3. Methods of conveying sick and wounded ashore. Women's Work during the War. 1. Uniforms worn by W.R.N.S. Officers and Ratings. 2. Personal Relics and Specimens. War Memorials. Photographs, Plans, elevations and Models of memorials erected to the memory of Navy, Army, and Air Force Officers and Men. Dioramas, e.g. of Zeebrugge. Photographic and Film Records illustrative of categories listed above.

The list on the left was submitted by the Imperial War Museum. The list on the right was submitted by the National Maritime Museum. Both are reproduced verbatim (The National Archives, EB 5/4).

thereof; visual media depicting offensive and defensive tactics and strategies; uniforms; ephemera; publications and medals—anything which helped roundly narrate the national and imperial war effort at sea. As Bradley's memorandum summarises, the museum wished to receive a 'balanced selection so that all aspects of the war effort of the Empire may be fully represented in the only Museum that can give a worthy picture of the whole'.[85]

The National Maritime Museum broadly supported the case of the Imperial War Museum in its own evidence submitted to the allocation subcommittee. At Callender's interview, he stated that 'the main object of the Imperial War Museum seemed to be to provide for the casual study of popular or spectacular war exhibits, many of which were connected with great events'.[86] The central tenets behind the argument put forward by the Imperial War Museum therefore can be seen to have held wider support. Yet a comment made in Callender's previously quoted oral testimony would influence the debate around its remit. This was use of the word 'spectacular' in describing the exhibits he felt the Imperial War Museum should be collecting versus those collected by the National Maritime Museum.

There are two references which Callender could have been making when he framed the collection of the Imperial War Museum as spectacular. The first concerns its stirring qualities. Even by 1939, many composite objects comprised large items of *matériel* such as complete or pieces from military vehicles, aircraft, boats and ships, or grand memorial paintings such as John Singer Sargent's *Gassed*, Paul Nash's *The Menin Road* and Percy Wyndham Lewis' *A Battery Shelled*, all which can induce shock and awe in visitors at the power they enable the state to wield, or the horrors this power can unleash when wielded. The second concerns a perception that its collection had little high cultural-historical value. After all, the Imperial War Museum was revolutionary in seeking to court popular appeal by presenting, interpreting and preserving ordinary objects.[87] He could also have been inferring both. In any case, his emphasis was on the influence they had amongst visitors.

Callender's motive for invoking the phrase 'spectacular war exhibits' is uncertain. While both possible reasons proposed above are appropriate to describe many items from the collection of the Imperial War Museum both then and now, at no point does Cubitt's letter, Bradley's oral testimony, or the memorandum describe any of the institution's existing or desired exhibits as possessing anything like spectacular attributes. Indeed,

Cornish's study on the museum's object collection and display rationales throughout 1917–1939 implies that spectacularness, as a consciously defined concept, did not feature in the curatorial decision making there before the Second World War commenced.[88] Obviously, this does not mean that before 1939 the museum tried to avoid collecting objects with arguably spectacular attributes. In another study on the museum's historical collecting practices, Cornish shows it clearly did so by collecting large, imposing items.[89] The reason for their collection however was not down to any spectacular attributes the objects may have had. Rather, it was for their provenance, the stories they helped convey about the lives of the people associated with them.[90] Similar findings are presented in research by Cundy.[91]

It is conjected that the motive behind Callender's intervention arose because he thought this dimension would help him to contrast his own institution's collecting remit with that of the Imperial War Museum. Otherwise, there may have arisen some crisis-conducive situation impacting the National Maritime Museum. Indeed, it is possible that Callender, justifiably or otherwise, perceived a duplication of effort discussed above arising between the Imperial War Museum and the National Maritime Museum, with all the ramifications this problem entailed for the legitimacy of the latter.[92] After all, judging from the Standing Commission's interim report on the various remits and claims, there clearly existed crossover between both: the Imperial War Museum being similarly interested in 'portraits, contemporary representations of big battles; ship models of chief ships; British Naval uniforms; naval ensigns; naval instruments; personal relics of very distinguished naval men [...]; library manuscripts and books, including Admiralty manuscripts up to the last war' which could convey stories through their provenance.[93] The spectacular, however, was something the National Maritime Museum 'would often not be interested in'.[94] It is possible that Callender thought by framing the collection of the Imperial War Museum as comprising little more than 'crowd pleasers' rather than supposedly serious historical items, he could ensure sufficient divergence between the two to avert duplication of effort over their public offer and, moreover, that the cream of what was available would end up at the National Maritime Museum. Supporting this idea, Callender also volunteered a list detailing what he thought would be the most appropriate items for the Imperial War Museum.[95] Presented down the right-hand column of Table 7.1, the emphasis, while demonstrating the same broad remit as the list forwarded by Bradley (see the left-hand column), was on so-called spectacular objects such as weapon delivery systems, ordinance and armour. Its implication is that high cultural-historical items should instead go to the National Maritime Museum.

The Outcome of the Report from the Material Allocation Subcommittee

This intervention by Callender is significant because it impacted the conceptual reinvention of the Imperial War Museum which, in turn, effected the institution's work following the Second World War. After considering all the evidence received from both the Imperial War Museum and National Maritime Museum, the Standing Commission's material allocation subcommittee recognised the latter as 'the premier establishment for the commemoration of the naval side of the present war'.[96] Its report therefore ruled that 'Records and relics of permanent historical interest which would continue the display of British maritime history on the lines of the existing collections should be assigned to the National Maritime Museum'.[97]

In contrast, the Imperial War Museum was accepted as an institution of popular appeal which catered for wider audiences. Accordingly, the Standing Commission acknowledged its claim to represent many and varied experiences and interests.[98] Yet, significantly, the report also agreed with the assertion made by Callender about the Imperial War Museum being concerned with collecting, displaying and interpreting 'spectacular' exhibits, going so far as reproducing the phrase.[99] The Standing Commission therefore adjudged that: 'The Imperial War Museum should be regarded as having prior claim to material with popular or spectacular appeal'.[100] No context was provided in the report for the term spectacular. As per Callender's original intervention however, its inference is on collecting and displaying star objects discussed above related to the sharp edge of the war at sea: for example, naval hardware, which the Standing Commission agreed that the National Maritime Museum would not typically have been concerned with displaying.

Developing the Nascent Conditions to Become a Museum of the Spectacular

Through the above process, the Standing Commission came to see and understand the Imperial War Museum as a museum which, alongside war work in everyday life, represented the shocking and awe-inspiring aspects of war. Laid before the Treasury on 14 August 1941, its report framed martial exploit and capability as an exhibit itself at the institution, comprising and upheld by individual items created from and for armed conflict. The production of this report comprises an important moment in the museum's history. While the institution reinvented from a museum on the 'war to end all war' to a museum on the two world wars, the discourse surrounding its mission started exploring new ideas. These ideas were eventually confirmed by publication during 1948. In its first public report for ten years on the national museums and galleries, the Standing Commission repeated the view that:

> the National Maritime Museum should be regarded as the premier establishment for the commemoration of the naval side of the war in so far as records and relics of personal historical interest were concerned, but that the Imperial War Museum should have a prior claim to material with a popular or spectacular appeal.[101]

No longer, therefore, was the Imperial War Museum being considered purely along its foundational lines. Instead, the Standing Commission legitimised greater interpretation of the characteristics of contemporary conflict which assimilated with perceptions over the role the Imperial War Museum now performed in light of the onset of the Second World War.

The above process broadened perspectives on the conceptual underpinnings of the Imperial War Museum. Before the Second World War, the museum was treated principally as commemorative. The discourse arising from the conflicting demands between the Imperial War Museum and the National Maritime Museum over naval material and, subsequently, the Standing Commission's report on material allocation, however, provided the impetus for a new exhibition context wherein the drama of war enabled by greater engagement with inherently dramatic material and concepts could become more central to the museum's interpretation. This point is supported in a broader context by Clive Gray, who explains that: 'The justifications for continuing with existing patterns of operation for developing new approaches to the sector depend upon how [...] multiple conflicting demands [(such as these)] are politically managed, and the grounds upon which legitimacy for the sector is based'.[102] So, while post-Second World War public demand may have prompted the institution's technological framework, this episode enabled that demand to be addressed.

This outcome demonstrates that not all museum reinvention derives entirely from internal action. Sometimes, the process can be influenced by external, intersecting forces. Similarly to most organisational change, externally influenced conceptual reinvention stems from dissatisfaction.[103] Specifically, it stems from senses of organisational dissatisfaction arising out of perceptions of competition. This takes place amid a context where the prevailing organisational philosophy is not predicated on outcompeting competitors, such as in the museum sector.[104] The situation transpires when the dissatisfied museum—in this case, the National Maritime Museum—causes the competitor museum to make desired changes for them. As a result, any competition between the two is eliminated on the dissatisfied museum's behalf at the competitor's expense. The process occurs through lobbying key influencers connected with the competitor, including their stakeholders. When it occurs, externally influenced conceptual reinvention has potential to impact a reinvention programme by influencing the conception of the vision which the transforming museum seeks.[105]

In the end, the report produced by the Standing Commission's exhibit allocation subcommittee addressed the question of distributing potential exhibits amongst the museums using 'broad principles'.[106] Through doing so, it defined loose categories for the pending material, and then indicated where the material comprising those categories should be deposited. But the nature of the material under each category was left open. Consequently, the report did not bring complete closure to the question over what objects should be deposited at which museums. It did however set the terms of the debate for when the debate eventually took place. To address this question, the report recommended that a disposals board be established which could 'decide the allocation of objects of value from an historical point of view as representing the share of the Service and other Departments in the war effort'.[107] Although its constitution was not dwelled on, the report did recommend that the board be composed of a representative from the three service departments, the Ministry of Home Security for civil defence matters and the three national museums.[108]

Conclusion

Before the Second World War, the Imperial War Museum, aside from embodying a museum on the 'war to end all war', also embodied a commemorative museum. It sanctified and memorialised the war dead of the

United Kingdom, British Empire and Dominions. Through doing so, the museum did not present technology for technology's sake. Rather, it presented objects with stories, many of these happening to be technological. Any comments about such attributes contributed towards the commemorative interpretative framework that the museum employed. Yet, in becoming a museum on the two world wars, the Imperial War Museum also became, gradually if not suddenly, a museum on war technology, with both technology and commemoration competing for prominence amongst the interpretation. As the post-war years advanced, the technological narrative won out. This shift occurred via the conceptual reinvention process, undertaken towards its institutional reinvention programme, responding to the crisis of irrelevancy raised by the onset of the Second World War.

Through this example from the case study, the current chapter argues that successful institutional reinvention in the museum context involves much more than the reinvention of what a museum does. It also involves the reinvention of what a museum is and is for: an essential consideration for the process to be meaningful. Without this understanding, museum reinvention risks taking place either without clear direction or without the project ever actually commencing, both outcomes potentially causing significant ramifications in contexts where organisational change has been deemed necessary. Indeed, only when possessing vision about some future desired state can actual, considered and effective museum reinvention confidently take place. At reaching this conclusion, the chapter also argues that conceptual reinvention is not always strictly an internal affair. It can also be influenced by external forces, especially where those forces derive from a vested interest in the outcome.

By making its case to the Standing Commission on Museums and Galleries for exhibits from the Second World War, the Imperial War Museum demonstrates being in possession of clear plans towards reinvention. The evidence submitted conveys a vision that would see the museum represent both world wars along existing interpretive frameworks and established representational and demographic lines. This vision would not materialise as originally conceived, however. The intervention of Sir Geoffrey Callender, acting for the National Maritime Museum, in the work by the Standing Commission on Museums and Galleries to determine what material should be deposited at which museum following the Second World War, catalysed an emergent, alternative discourse about the Imperial War Museum. Specifically, it disrupted the received wisdom on what the museum was and was for. This unexpectedly modified the institution's conceptual reinvention. After the Second World War, Callender's intervention inadvertently enabled the Imperial War Museum, through its reinvention, to respond to post-war public interest in technology, with the relevant issues and ideas having been factored into the museum's wartime reconceptualisation.

By focussing on the conceptual reinvention of the Imperial War Museum, this chapter has considered the first stage in the process comprising museum reinvention. The second is translating reinvention. This transpires when a museum instigates new practices and states of existence in line with its newly reconceived conceptual underpinnings. The next chapter explores that second process, focussing on the museum's first steps to acquiring new collection material and storage space so the anticipated new material could be received.

Notes

1. The National Archives (TNA), EB 5/4, 'Disposal of War Relics: Summary Note on the Present Position', October 1940, 2.
2. Ibid., 3.
3. TNA, EB 2/1, 'Note of Interview with Mr. L.R. Bradley, Director of the Imperial War Museum', 2.
4. Gaynor Kavanagh, *Museums and the First World War: A Social History* (London: Leicester University Press, 1994), 157.
5. Catherine Pearson, *Museums in the Second World War: Curators, Culture and Change*, ed Suzanne Keene (Abingdon: Routledge, 2017), 32.
6. IWM, Museum Archive (MA), EN1/1/MUSG/013, Wardley to Bradley, 6 March 1934.
7. Willie Pietersen, *Reinventing Strategy: Using Strategic Learning to Create and Sustain Breakthrough Performance* (New York, NY: John Wiley).
8. Christine M. Pearson and Ian I. Mitroff, 'From Crisis Prone to Crisis Prepared: A Framework for Crisis Management', *The Executive* 7 (1993): 52–53.
9. David Fleming, 'The Essence of the Museum: Mission, Values, Vision', in *Museum Practice*, ed Conal McCarthy, The International Handbooks of Museum Studies (Chichester: Wiley Blackwell, 2015), 3.
10. Gail Anderson, 'A Framework: Reinventing the Museum', in *Reinventing the Museum: The Evolving Conversation on the Paradigm Shift*, ed Gail Anderson (Lanham, MD: AltaMira, 2012), 2.
11. Richard Beckhard, 'Strategies for Large System Change', *Sloan Management Review* 16 (1975): 45.
12. Ibid.

13. Jonathan Law, *A Dictionary of Business and Management*, 6th edn (Oxford: Oxford University Press, 2016), s.v. 'situational analysis'.
14. Pietersen, *Reinventing Strategy*, 71.
15. Gary Edson and David Dean, *The Handbook for Museums* (London: Routledge, 1996), 27–28.
16. Pietersen, *Reinventing Strategy*, 106.
17. Ibid.
18. John Falk, 'The Visitor Museum Experience: Who Visits, Why and to What Effect?', in *Reinventing the Museum: The Evolving Conversation on the Paradigm Shift*, ed Gail Anderson (Lanham, MD: AltaMira, 2012), 325–327.
19. Carol M. Komatsuka, 'Expanding the Museum audience Through Visitor Research', in *Museum Management and Marketing*, ed Richard Sandell and Robert R. Janes (London: Routledge, 2007), 373.
20. Pietersen, *Reinventing Strategy*, 115–117.
21. Fiona McLean, *Marketing the Museum* (London: Routledge, 1997), 117.
22. International Council of Museums, *Code of Ethics for Museums* (Paris: International Council of Museums [2004?], 20–21, https://icom.museum/wp-content/uploads/2018/07/ICOM-code-En-web.pdf (18 March 2024).
23. G. Brown Goode, 'The Principles of Museum Administration', in *Museum Provision and Professionalism*, ed Gaynor Kavanagh (London: Routledge, 1994) 41–46; Alf Hatton, 'Current Issues in Museum Training In the UK', in *Museum Management*, ed Kevin Moore (London: Routledge, 1994), 148–155; H. L. Madison, 'Tentative Code of Museum Ethics', in *Museum Provision and Professionalism*, ed Gaynor Kavanagh (London: Routledge, 1994), 265–270; J. Kenyon, 'Collecting for the Twenty-First Century', in *Collections Management*, ed Anne Fahy (London: Routledge, 1995), 122–125.
24. Goode, 'Principles, 46; Kenyon, 'Collecting', 122–123.
25. Paul Cornish, 'Extremes of Collecting at the Imperial War Museum, 1917–2009: Struggles with the Large and the Ephemeral', in *Extreme Collecting: Challenging Practices for 21st Century Museums*, ed Graeme Were and J. C. H. King (New York, NY: Berghahn, 2012), 162.
26. Robin Luckham, 'Armament Culture', *Alternatives* 10 (1984): 4–5.
27. Theodor W. Adorno, 'Education after Auschwitz', trans. Henry W. Pickford, in *Can One Live After Auschwitz? A Philosophical Reader*, ed Rolf Tiedemann (Stanford, CA: Stanford University Press, 2003), 29.
28. Ibid.
29. Robert Bevan, *The Destruction of Memory: Architecture at War*, 2nd edn (London: Reaktion, 2016), 17–18; see Philip Warner, *Firepower: From Slings to Star Wars* (London: Grafton, 1988).

30. Cornish, 'Extremes of Collecting', 162.
31. Adorno, 'Education after Auschwitz', 29.
32. Ibid.
33. Charity M. Counts, 'Spectacular Design in Museum Exhibitions', *Curator* 52 (2009): 274.
34. Guy Debord, *Society of the Spectacle* (London: Rebel/Dark Star, 1977); Anne McClintock, *Imperial Leather: Race, Gender and Sexuality in the Colonial Contest* (New York, NY: Routledge, 1995), 56–61; Bennett, *The Birth of the Museum*, 63–69; Nick Prior, 'Postmodern Restructurings', in *A Companion to Museum Studies*, ed Sharon Macdonald (Maldon, MA: Blackwell, 2006), 514.
35. Paul Greenhalgh, 'Education, Entertainment and Politics: Lessons from the Great International Exhibitions', in *The New Museology*, ed Peter Vergo (London: Reaktion, 1989), 74–98.
36. Kasia Tomasiewicz, '"We are a social history, not a military history museum"': Large Objects and the "Peopling" of Galleries in the Imperial War Museum, London', in *Museums, Modernity and Conflict: Museums and Collections in and of War Since the Nineteenth Century*, ed Kate Hill (Abingdon: Routledge, 2021), 222. The author has also encountered and noted numerous pieces of anecdotal evidence informing this point through interactions with curators and other staff while working and researching at IWM.
37. Francesca Monti and Suzanne Keene, *Museums and Silent Objects: Designing Effective Exhibitions* (Abingdon: Routledge, 2016), 271.
38. James Scott, 'Objects and the Representation of War in Military Museums', *Museum and Society* 13 (2015): 495–497.
39. Alys Cundy, 'Objects of War: The Response of the Imperial War Museum, London, to the First and the Second World Wars', *Post-Medieval Archaeology* 51 (2017): 268–269.
40. Ibid., 269.
41. Cornish, 'Extremes of Collecting', 161–162.
42. Imperial War Museum, *A Short Guide to the Imperial War Museum*, 3rd edn ([London?]: [HMSO?], [1946?]), 2.
43. H. Shaw, 'Science Museums', *Museums Journal* 46 (1946): 173.
44. Adorno, 'Education after Auschwitz', 29.
45. W. Richard Scott, *Institutions and Organizations: Ideas, Interests and Identities*, 4th edn (Thousand Oakes, CA: SAGE, 2014), 72.
46. Maria Bonnafous-Boucher and Jacob Dahl Rendtorff, *Stakeholder Theory: A Model for Strategic Management* ([Cham?]: Springer, 2016), 2.
47. TNA, EB 3/19, Bradley to First Lord of the Admiralty, 10 September 1939, 1.
48. TNA, EB 3/19, anon. to the Secretary of the Admiralty, 31 October 1939.

49. TNA, EB 5/4, Callender to Beresford, 30 April 1940, 1.
50. TNA, EB 3/19, Callender to the Secretary of the Admiralty, 13 December 1939, 1.
51. Ibid.
52. TNA, 3/19, Callender to the Secretary of the Admiralty, 23 January 1940, 1–2.
53. Ibid., 1.
54. TNA, EB 2/1, 'Note of Interview with Mr. L.R. Bradley, Director of the Imperial War Museum', 1.
55. IWM, MA, EN2/1/COB/049, Mackintosh to Bradley, 17 January 1940, 1.
56. Nigel Rees, *Don't You Know There's a War On? Words and Phrases from the World Wars* (London: Batsford, 2011), s.v. 'Nazi'.
57. TNA, EB 3/19, Beresford to the Trustees of the Imperial War Museum, 23 April 1940, 1.
58. TNA, EB 2/2, 'Sub-Committee on the Allocation of Relics of the Present War', 2.
59. TNA, 3/19, Callender to Beresford, 24 April 1940.
60. TNA, EB 5/4, 'Disposal of War Relics: Summary Note on the Present Position', 1.
61. Brian Foss, *War Paint: Art, War, State and Identity in Britain, 1939-1945* (New Haven, CT: Yale University Press, 2007), 17–31.
62. TNA, EB 2/2, 'Sub-Committee on the Allocation of Relics of the Present War: Report', 1.
63. TNA, EB 2/2, Standing Commission on Museums and Galleries, 14 August 1941, 3.
64. TNA, EB 2/2, 'Sub-Committee on the Allocation of Relics of the Present War: Report', 5.
65. Ibid.
66. Ibid.
67. Ibid.
68. Timothy Ambrose and Crispin Paine, *Museum Basics: The International Handbook*, 4th edn (Routledge: Abingdon, 2018), 381–390.
69. Timothy Ambrose, 'The Purpose of a Forward Plan: Solving Problems; Making the Best Use of Resources; Getting Value for Money', in *Forward Planning: A Handbook for Business, Corporate and Development Planning for Museums and Galleries*, ed Timothy Ambrose and Sue Runyard (London: Routledge, 1991), 2–5.
70. TNA, EB 5/4, Cubitt to Beresford, 25 April 1940.
71. TNA, EB 5/4, 'The Case for Including in the Imperial War Museum Relics, Trophies, and Records of the Present War in All Its Aspects, at Sea,

on Land, in the Air on the "Home Front", Together with Those of the War of 1914–1918'.

72. TNA, EB 2/1, 'Note of Interview with Mr. L.R. Bradley, Director of the Imperial War Museum'.
73. TNA, EB 5/4, 'A Short List of Classes of Naval Material Required for the Imperial War Museum Based on the Contents of the Present Naval Galleries and the Reference Departments'.
74. Marianna Adams and Judy Koke, 'Comprehensive Interpretive Plans: A Framework of Questions', *Journal of Museum Education* 33 (2008): 293–300; Pietersen, *Reinventing Strategy*, 69–104.
75. John P. Kotter, 'Leading Change: Why Transformation Efforts Fail', in *Museum Management and Marketing*, ed Richard Sandell and Robert R. Janes (Abingdon: Routledge, 2007), 27; Pietersen, *Reinventing Strategy*, 105–126.
76. Beckhard, 'Large System Change', 45.
77. Kavanagh, *Museums*, 137; Sue Malvern, 'War, Memory and Museums: Art and Artefact in the Imperial War Museum', *History Workshop Journal* 49 (2000): 185–188.
78. TNA, EB 5/4, Cubitt to Beresford, 25 April 1940, 1.
79. TNA, EB 5/4, 'The Case for Including in the Imperial War Museum Relics, Trophies, and Records of the Present War in All Its Aspects, at Sea, on Land, in the Air on the "Home Front", Together with Those of the War of 1914–1918', page 3 of the document (page 5 of the overall submission).
80. Ibid., 6.
81. Jennifer Wellington, *Exhibiting War: The Great War, Museums, and Memory in Britain, Canada, and Australia* (Cambridge: Cambridge University Press, 2017), 242.
82. Alys Cundy, 'The Imperial War Museum and the Material Culture of the First World War, 1917–2014', in *The Edinburgh Companion to the First World War*, ed Ann-Marie Einhaus and Katherine Isobel Baxter (Edinburgh: Edinburgh University Press, 2017), 406.
83. TNA, EB 2/1, 'Note of Interview with Mr. L. R. Bradley, Director of the Imperial War Museum', 1.
84. Cundy, 'Objects of War', 265–266.
85. TNA, EB 5/4, 'The Case for Including in the Imperial War Museum Relics, Trophies, and Records of the Present War in All Its Aspects, at Sea, on Land, in the Air on the "Home Front", Together with Those of the War of 1914–1918', page 5 of the document (page 7 of the overall submission).
86. TNA, EB 5/4, 'Disposal of War Relics: Summary Note on the Present Position', 2.

87. Cundy, 'Imperial War Museum', 406.
88. Paul Cornish, '"Sacred Relics": Objects in the Imperial War Museum, 1917–1939', in *Matters of Conflict: Material Culture, Memory and the First World War*, ed Nicholas J. Saunders (Abingdon: Routledge, 2004), 35–50.
89. Cornish, 'Extremes of Collecting', 157–167.
90. Cornish, '"Sacred Relics"', 37.
91. Alys Cundy, 'Thresholds of Memory: Representing Function through Space and Object at the Imperial War Museum, London, 1918–2014', *Museum History Journal* 8 (2015): 247–268; Cundy, 'Objects of War', 261–273.
92. Goode, 'Principles', 46; Kenyon, 'Collecting', 122–123.
93. TNA, EB 5/4, 'Disposal of War Relics: Summary Note on the Present Position', 2.
94. Ibid.
95. TNA, EB 5/4, 'Categories Suitable for the Imperial War Museum'.
96. TNA, EB 2/2, 'Sub-Committee on the Allocation of Relics of the Present War', 3.
97. Ibid., 4.
98. Ibid.
99. Ibid.
100. Ibid.
101. Standing Commission on Museums and Galleries, *Third Report: The War Years and After* (London: HMSO, 1948), 18.
102. Clive Gray, *The Politics of Museums* (Basingstoke: Palgrave Macmillan, 2015), 22–23.
103. Beckhard, 'Large System Change', 45.
104. Kenyon, 'Collecting', 122–125; International Council of Museums, *Code of Ethics.*
105. Beckhard, 'Large System Change', 45.
106. TNA, EB 2/2, 'Sub-Committee on the Allocation of Relics of the Present War', 6.
107. Ibid.
108. Ibid.

CHAPTER 8

Materialising the Reconception, 1940–1948

Introduction

The current chapter is the last case study chapter herein and, thus, also the last chapter to explore the crisis of irrelevancy at the Imperial War Museum during the Second World War. Previously, Chap. 6 contextualised this crisis-conducive situation confronting the museum. It explored how the museum detected the signals, sensed the significance of the problem and conceived the necessary resolution and mitigating strategy—an institutional reinvention, bringing about a new *raison d'être* and rationale for the museum. Chapter 7 then considered the museum's conceptual reinvention in response to the crisis. It explored a multi-stakeholder, internal and external process which established the museum's post-war subject remit and exhibition priorities. And now, finally, the current chapter considers actions of the museum to deliver on that conceptual reinvention. It explores several key actions which ensured the conceptual reinvention was realised in the museum's organisational elements. Through doing so, the chapter also explores different challenges underlying any strategy by which museum reinvention is realised. It contends that institutional reinvention processes require patience, creativity and capacity for adjustment once underway.

The historical context of the current chapter is the privations and national strategic demands of British society during the years 1940–1948 of the wider Second World War era. This context arose from the total war effort which the conflict had transitioned into in the United Kingdom

P. W. Deans, *Crisis, Reinvention and Resilience in Museums*, https://doi.org/10.1007/978-3-031-67806-6_8

from May 1940. All persons and property became liable to the compulsion of the state. Indeed, everything useful became either a resource for societal subsistence or military application as well as an enemy target.[1] Alongside the death and destruction caused by the hostilities, mass rationing and resource prioritisation also arose to hamper everyday processes and operations, limit nonessential projects and generally making regular consumptive activities less achievable and permissible. These privations and restrictions dogged the Imperial War Museum throughout the Second World War and well into the post war years. Accordingly, the museum faced many challenges when attempting to realise its reinvention, requiring multiple methods and changes to the task at hand.

The current chapter continues presenting the reinvention of the Imperial War Museum using the two-stage articulation established in Chap. 7 which draws on ideas forwarded by reinvention theorist Willie Pietersen.[2] Whereas the first stage, explored over the previous chapter, involves conceiving a new *raison d'être* and rationale for the museum, the second stage, explored below, concerns actions to materialise this new paradigm through reconfiguring the museum's various organisational elements. The current chapter considers what is framed here as 'translating reinvention'. It comprises work geared towards analysing and understanding the outcome of the conceptual reinvention process and interpreting how that vision can be realised. In line with Chap. 7, the current chapter frames this activity as constituting a preventative crisis management strategy, which, for context, is mapped against Christine M. Pearson and Ian I. Mitroff's general five-stage crisis management framework.[3] The chapter also continues using David Gleicher's reinvention formula to inform understanding of the structure and nature of the museum's reinvention.[4]

The current chapter addresses the above over three substantive sections. The first section explicates the translating reinvention process. It elucidates the concept and any relevant associated issues, ideas and activities. The second section then considers two different activities which the Imperial War Museum performed to translate conceptual reinvention. It first explores how this translation was achieved against the museum's collection. It then explores how the translation was achieved against the museum's spatial facilities. And the third section discusses how the threat posed by cultural irrelevancy and the counter measures which the museum deployed in response comprise an example of deep-revolutionary crisis and deep revolutionary crisis management. It synthesises ideas from the concept with findings from the case study presented over Chaps. 6, 7 and 8.

Translating Reinvention

If the conceptual reinvention of a museum concerns renewing the museum's *raison d'être* and rationale, translating reinvention can be understood as interpreting conceptual reinvention and applying it against that museum's organisational elements. Generally, translation involves taking meaning mediated through one form and conveying that meaning accurately through another.[5] In the institutional reinvention context therefore, translation involves taking the meaning for some new proposed paradigm mediated through established plans and conveying that meaning through the various composite elements of the organisation, accounting for modifications or radical changes where needed.

At first glance, translating reinvention might appear little more than delivering on some museum's conceptual reinvention. This is because, at face value, the work of conceptual reinvention comprises a series of complex and creative thought processes that define how some institutional reinvention should transpire, whereas the work of translating reinvention comprises merely a series of practical actions which serve to make institutional reinvention happen. But such an understanding underrates its cognitive complexity. After all, translating reinvention is more than just a process of physically delivering on conceptual reinvention. It is also a process of working out how to deliver the reinvention. Translating reinvention involves comprehending the new *raison d'être* and rationale conceived for some museum, determining the existing structure comprising that museum and the elements therein which need to be modified so the museum can cohere with the new paradigm, and implementing those structural alterations as per the stated requirements.

The process of translating conceptual reinvention onto a museum involves reconfiguring its manifold organisational elements in line with the desired state. Pietersen thematises the various issues, ideas and activities which comprise this work as 'alignment'.[6] It involves readying the museum by preparing those elements to support the change. Otherwise, a museum will be incapable of assuming the new paradigm. Pietersen proposes four organisational elements that must be prioritised for successful reinvention: measures and rewards, structure and process, culture, and people. The first element, measures and rewards, concerns what defines and drives organisational success.[7] When reinventing a museum, recalibrating the way outcomes are gauged and the nature of the incentives which encourage them may be necessary. This is because through adopting an alternative paradigm, the pre-existing methodology for defining and driving success might be rendered obsolete. For example, the museum that reinvents from a for-profit to a more typical not-for-profit model may find its former commercial performance measures and rewards unpracticable in the third sector and therefore needing replacement, and *vice versa*.[8] The second element, structure and process, concerns physical and conceptual aspects that constitute organisations: facilities, departments, teams, output, *etcetera*.[9] Depending on the outcome of a museum's conceptual reinvention, such aspects may need modification to facilitate the new *raison d'être* and rationale.[10] This could be by changing the way departments are formed, reconstituting management structure or amending the physical assets available as required.[11] The third element, culture, concerns ideas, customs and transactions of staff in an organisation.[12] Specifically, these are the routine or predictable behaviours derived from collective or dominant understandings, perceptions and assumptions. Aligning organisational culture with the new paradigm is essential for successful museum reinvention.[13] An unsupportive culture may deliberately or unintentionally undermine a new strategy, threatening its sustainability.[14] And the fourth element, people, concerns workforce.[15] Museums rely on people for their continuing operation. It is important therefore that staff are brought about to see the need for reinvention and can be focussed on making the necessary interventions.[16] Dissatisfied staff, or staff fearful of what change might mean, may consciously or otherwise prevent museum reinvention from occurring.[17] All this demonstrates that an alignment of organisational elements must include not just physical elements, but non-physical elements too.

As well as reconfiguring organisational elements, the process of translating reinvention onto a museum involves overseeing the process's successful completion. Pietersen thematises the various issues, ideas and activities which comprise this work as 'execution'.[18] In doing so, he warns against framing it as the final implementation of the reinvention process. This is because the whole change process, all the activities associated with conceptualising and translating reinvention, go towards delivering on the implementation. Rather, Pietersen frames execution as a reinvention's successful delivery through reflexive learning.[19] Put another way, execution is the work of reflexively adjusting the process where challenges arise to bring about the desired outcome. There are two core considerations which should be accounted for in this regard. The first is the environmental context (defined broadly) wherein a museum's reinvention takes place. Those planned and carried out under ideal circumstances obviously stand greater chance of being easily and cleanly implemented compared with those which are poorly planned and/or carried out under crisis conditions. Yet, most institutional reinventions rarely occur, if ever, under ideal circumstances. Only when the urgent imperative for change arises, with all the haste and trauma this situation can necessitate and entail, do museums usually find themselves embarking on reinvention.[20] Moreover, in the messy, unpredictable real world, external and internal events can transpire which disrupt even the most carefully orchestrated reinvention processes by altering the dynamics of the situation. This may require abandoning certain initiated interventions, or require new interventions to be plotted.[21] Reinvention processes therefore cannot be taken for granted. Their execution is often unsystematic and difficult, a reality which museums contemplating reinvention should fully anticipate. The second consideration concerns the need for an ability and willingness to experiment during the organisational reinvention process.[22] If a reinvention's environmental context has unpredictability or there are few resources available with which the reinvention can be delivered, the adopted change plan may itself require changes, perhaps multiple changes. Under these circumstances, the review of progress against the prevailing hypotheses on the cause-and-effect relationships underpinning the chosen strategy is vital. Where planned interventions become un-implementable or fail to produce a sought-after effect, the strategy should be modified until the desired state becomes fully instituted with all objectives met.[23]

Given the broad purview that translating reinvention entails, this chapter focusses on what are determined herein as the most important 'first steps' undertaken by the Imperial War Museum towards realising its reinvention.[24] These were, firstly, the steps to acquire new exhibits deriving from the Second World War and, secondly, the steps to acquire space with which to store those new exhibits. Without either, the reinvention could not have taken place at all: limited storage would have hampered the Imperial War Museum from developing its collection, while a limited collection would, in turn, have hampered the museum from attaining the vision set during the conceptual reinvention. These steps are therefore treated here as critical components towards the museum's transformation. In doing so, the chapter concentrates on the alignment of structures and processes at the Imperial War Museum: the most germane elements to both examples. It must also be emphasised that the chapter does not offer a complete exploration of these first steps. The sources on each are voluminous and events too numerous. Consequently, these first steps are discussed here with a focus on the key points in the processes, leaving broader considerations for another study.

Resolving a Crisis of Irrelevancy at the Imperial War Museum (III): Executing the Response

First Steps to Aligning the Collection

Aligning the collection at the Imperial War Museum is a clear example of what Pietersen calls 'structural alignment', interpreted here as including both physical and non-physical organisational elements.[25] It also comprised an appropriate response to the question which Pietersen recommends that all organisations should ask themselves when undergoing reinvention: 'To best support the new strategy, should the firm be organized by product line, customer grouping, function, geography, or some other principle?'[26] As demonstrated in Chap. 3, museum collections, whether tangible or intangible, underpin museums and museum purpose.[27] They are an important basis for the public support which uphold museums and thus become a (if not *the*) nucleus around which museums develop.[28] Indeed, the threatened onset of cultural irrelevancy brought about by the Second World War could only have been resolved if the collection had been reorganised as required. It follows therefore that the collection was the 'principle' around which the museum reinvented.

First steps by the Imperial War Museum towards acquiring new exhibits from the Second World War began once the Treasury had consented to its redefinition, discussed during Chap. 6. Most of those objects collected over the war years comprised two-dimensional records such as official, ephemeral and grey documentation. On occasion however, three-dimensional objects were also collected where the opportunity arose. When this occured, the benefactor tended to be a private individual or group.[29]

There were two reasons for why the Imperial War Museum started collecting with a focus on documents. The first was to minimise and hopefully prevent gaps from occurring in the museum's collection of ephemera. Although often replaceable, such material usually had short utilitarian lifespans. For example, while multiple, indeed potentially many copies of posters, pamphlets, leaflets, *etcetera* were printed in individual production runs, the short period wherein they tended to be used and/or their restrictive geographical distribution limited opportunities for their collection.[30] The second and more practical reason was because the Imperial War Museum 'could hardly concern ourselves with more',[31] as explained the Director-General, Leslie Bradley, at the war's outset. Aside from possessing limited space, the museum's building was an inappropriate venue for storing rare material because of its vulnerability discussed in Chap. 5. As Bradley went on to explain:

> One difficulty that we shall have to consider is that of storing the material during the war, for this building, as you know, is entirely unsafe, both in location and construction, and it was for this reason that we evacuated most of our valuable possessions, so that it would seem unwise to begin to full [*sic*] it again with new material, although of course, this does not apply to photographs so long as the negatives are kept in a safe place elsewhere.[32]

His comment here was made eleven months before the start of the Blitz, but it foreshadowed the problems faced by the museum during the bombings, discussed in Chap. 5.

This situation exemplifies the impact that environmental factors can have on translating reinvention and the need for reflexive behaviour in executing chosen strategies.[33] The context wherein change takes place will greatly influence their pace and course.[34] In the case of the Imperial War Museum, the context threatened to literally destroy any progress made towards the reinvention. This means that the translation required due consideration for both the strategy's progress and its success.[35] Focussing on the collecting of documents comprised a way forward in that regard. It enabled the museum to progress with a chance at repeat in work if need arose. Notwithstanding their limited print runs, the fact that posters, pamphlets, leaflets *etcetera* were produced in numbers mitigated somewhat the risk of their loss should the museum be significantly damaged. Though, this risk did not stop the museum accepting the original document representing the Joint Declaration signed by Neville Chamberlain and Adolf Hitler, mentioned in Chap. 4—an opportunity too good to miss.[36]

The Imperial War Museum sought for systematic collections to be made by national government, the local authorities and colonial administrations overseas during the Second World War. Collections of material from the national government were arranged via the Treasury. It sent letters to individual government departments requesting that the Imperial War Museum be included on their circulation lists.[37] Collections of material from local authorities, in contrast, were arranged directly with the authorities which produced it.[38] And collections of material from colonial administrations were arranged via the Colonial Office. This came about after the Colonial Office sent a memorandum by the Imperial War Museum encouraging administrators to organise collections which represented the military and civilian war effort in each territory.[39] Although the memorandum did include suggestions for likely three-dimensional material of interest, it still predominantly focussed on documentation.[40]

Notwithstanding occasions where individuals gifted three-dimensional items to the Imperial War Museum, work towards acquiring such exhibits began most earnestly near the war's conclusion. As discussed in Chap. 7, during 1940–1941, the Standing Commission on Museums and Galleries produced a report commissioned by the Treasury making recommendations for the distribution of material from the Second World War between the Imperial War Museum, National Maritime Museum, Science Museum and the Royal United Service Institute Museum.[41] One recommendation advised establishing a Disposals Board to ensure the dispersal formula it had created was applied fairly.[42] The Standing Commission submitted this report in August 1941, which was approved by the Treasury the following year.[43] But no further action occurred until April 1945, when a promptive letter from the Royal United Service Institute Museum stimulated the Standing Commission to resurvey each museum on whether its recommendations still held relevance.[44]

The Trustees of the Imperial War Museum considered this question at their meeting on 8 May 1945. Through doing so, they 'reaffirmed their view that it was for the Departments to give recommendations and advice to the Trustees as to what exhibits should be included in the Museum to represent the Department's share in the war effort'.[45] Reporting this decision back to the Standing Commission, Bradley contextualised their position by emphasising the information requirements and support they needed in overseeing such work:

> [I]t would be quite impossible for the Trustees to be in possession of sufficient information to enable them merely to ask the services for this or that particular item with any hope that the resulting collections would form an adequate record of the material used by the Services and Home Front Departments during the war.[46]

With the other three museums also in favour of convening the Disposals Board, a meeting about the proposal was held on 17 December 1945 attended by representatives from those museums, including Bradley, and the relevant government departments concerned.

The issue behind this position stemmed from the diminished ability of the Imperial War Museum to collect material compared with its ability during and following the First World War. Since 1918, potential acquisitions had dramatically increased in size. Indeed, Bradley had previously concluded that, through the 'tendency since 1914-1918 [...] towards increased size of material', any collection of the new war would have to be 'much less ambitious and very much more selective'.[47] He further justified this point by observing that 'guns are certainly no smaller, types of tanks and mechanized vehicles more numerous'.[48] This recognition marks what Pietersen calls 'processual alignment'.[49] Specifically, it was an alignment of the process by which the museum collected exhibits. As Pietersen explains: 'Moving to a more innovative mode will probably require significant changes in [...] how [...] decisions get made'.[50] Where, following its foundation, the museum had been concerned with making a collection of ethnographic proportion, the environment created by the Second World War forced the museum to follow this new conflict with philosophy and procedures that made the museum more discriminatory over what would and would not be collected.[51]

To call the meeting on 17 December 1945 the Disposals Board, as other scholarship has done, would be a misnomer.[52] No actual dispersals resulted from it. This is partly because the progress by the service departments on earmarking objects for the museums had occurred unevenly: a further nod at the environmental challenges which can impact on museum reinvention.[53] In line with pre-Second World War policies and activities, the Air Ministry had actively pushed for the Royal Air Force to maintain an active earmarking programme[54]—albeit with the vision of eventually establishing its own aviation museum.[55] The War Office, by contrast, had done the opposite. Five years of intense pressure on the British Army meant very few land-based related items were earmarked over the Second World War.[56] Occupying a somewhat intermediary position, the Admiralty had sought the Royal Navy to make lists of material that could be earmarked if requested by the museums concerned.[57]

It is worth reflecting on the problems this disparity posed the reinvention of the Imperial War Museum. At their combined peak strength in June 1945, the British Army and the Women's Auxiliary Territorial Service comprised some 3,110,800 service personnel. This contrasted greatly with the relevant services of the Admiralty and Air Ministry. The naval services had peaked at 855,000 during June 1945 and the air services had peaked at 1,176,400 in June 1944.[58] For an institution that reputedly served the everyday person, providing insufficient representation of the experience of the services which had employed the greatest number of individuals could have been greatly delegitimising. Indeed, it risked entrenching the very crisis which the reinvention had been initiated to avert in the first place.

Despite these pitfalls, the meeting of 17 December achieved progress towards the intended outcome. It was agreed that the Admiralty and Air Ministry would immediately circulate documents listing the available material amongst the relevant museums. The War Office would also prepare lists of possible material and circulate.[59] From these documents, the museums would indicate *prima facie* what they wanted. If any disputes arose, the matter would be referred to the Standing Commission for a decision. Where no decision was attainable, the Treasury would arbitrate.[60] To avoid this possibility however, the museum directors agreed that a meeting should be held between them before requests were made, so they could clarify each other's aspirations.[61] Figures 8.1 and 8.2 depict various

Fig. 8.1 Photograph depicting a Spitfire, jet engine and rocket projector on display, c. 1946. © IWM (D 29424)

Fig. 8.2 Photograph depicting a selection of 'Enemy Relics' on display, c. 1946. © IWM (D 29423)

examples of material which the Imperial War Museum obtained through this process, displayed just after the war when the museum reopened with its own exhibition on 26 November 1946.[62]

First Steps to Spatial Alignment

Arranging to acquire three-dimensional material from the service departments was necessary for the reinvention of the Imperial War Museum. But their actual acquisition would be impossible if the museum did not acquire the requisite storage space. Early recognition of this requirement prompted the museum to begin the first steps of aligning its spatial requirement with the newly conceived paradigm during July 1940, before the precise nature of the institution's reinvention discussed in Chap. 7 had been fully conceived. A second structural alignment process therefore began via an initial rationalisation programme, which was undertaken sporadically until 1942.[63] It resulted in collection items with 'no particular history', or 'the possibility of exhibiting which seemed very remote', being scrapped.[64] Rationalisations occur when museums want to reduce the size of or streamline their collection. This could be for various reasons. It may have become unmanageable. It could possess too many duplications. It may even no longer represent or support the mission, values and vision of a museum.[65] Whatever the reason, the rationale for any rationalisation must be one that supports the aims and objectives driving the museum in question.

The material sent for scrap comprised various items from the collection of the Imperial War Museum. These included various different guns, shells, aeroplane engines and naval equipment: objects made from industrial grade metals.[66] Today, collection rationalisations are governed in most countries by strict policies and procedures predicated on law, ethics and other standards.[67] Whether ethical considerations factored into the decision at the Imperial War Museum during the Second World War remains elusive. One could argue however, that given the pressing need for space, as well as the urgent public call by the government around that time for useful scrap materials, the circumstances afforded the decision some legitimacy.[68] Either way, it was undertaken within the parameters of the law governing the museum, because clause 2c of the *Imperial War Museum Act* 1920 afforded the museum the right to 'dispose of any objects belonging to the Museum which the Board consider unfit to be preserved or not to be required for the purposes thereof'.

This initiative successfully freed up some space at the Imperial War Museum. Yet the results were paltry considering the amount required; it did not generate enough display or storage space for a full implementation of the proposed reinvention. Considerably more would be needed than could have been achieved through rationalising the existing collection alone. This meant further approaches to aligning the museum's structure were necessary.

The next method involved redeveloping its building. This initiative made sense in principle, considering that the whole reinvention process was being undertaken to ensure the museum's long-term continuance. An enlarged building would, after all, expand both the storage and display space available there. Discussion with the Ministry of Works regarding this redevelopment was instigated by the Director-General, Bradley, during

February 1943, which resulted in government architects drawing up sketch plans for possible developments.[69]

The first sketch plans were shared with the staff and Trustees of the Imperial War Museum on 1 November 1943. This scheme utilised all four floors in the building and made provision for two visitor lifts, a cinema, workshops and library stacking that could hold 140,000 books and other such material.[70] To achieve it with minimal potential disruption of the museum, the architects cleverly drew up the plans so that the proposed development could be delivered incrementally over time as and when resources and manpower became available.[71] Two subsequent sketch plans were submitted in 1944. The second scheme responded to a realisation by the architects at the Ministry of Works that more modern architectural techniques could elicit further space than previously realised.[72] The third scheme then responded to concerns from the museum's staff and Trustees about the utility of certain floors and facilities as presented in the second sketch plan.[73] By May 1944 therefore, various changes had been proposed following the first plans. These included provision of over 100,000 square feet of space, at least 57,000 more than before the war; the placing of all the art galleries on the top floor; and the excavation of the ground floor.[74]

During transactions with the Ministry of Works regarding the proposed redevelopment of the Imperial War Museum, both Bradley and the Board of Trustees were repeatedly informed that these sketch plans should be considered as nothing more than that.[75] They comprised ideas which represented what could conceivably be possible on the existing site. The restoration of homes and essential facilities after the war in the immediate post-war period however would ultimately determine the speed at which this development could take place. Accordingly, only several months after the architects had produced their third sketch plan, Bradley, having realised the unlikeliness of any development to the museum occurring soon, was already discussing a third method with the Ministry of Works.[76]

This diverse approach demonstrates Pietersen's assertion that organisations must be prepared to experiment with and alter their reinvention strategy as required.[77] In fluid environments such as that experienced by the Imperial War Museum during the Second World War era, an inflexible change programme can put successful reinvention at risk. While a given strategy may once have suited an environment, physical and ideological changes therein may render the strategy unsuitable going forward.[78] Museums undergoing reinvention therefore must be prepared to reflexively modify their reinvention strategy. Academic futurist Paul Saffo describes this as the 'ready, fire, steer' approach of organisational development.[79] It occurs where some new direction is conceptualised and an implementation plan established and executed, all the while accounting for and accommodating unforeseen events or challenges. This includes 'ditching and switching' approaches as required.

The third method, and the one which would eventually resolve this problem faced by the Imperial War Museum, involved acquiring external storage space. It comprised a long and arduous if ultimately successful process for the museum, lasting approximately four years from mid-1944 until mid-1948. The earliest discovered document making significant reference to the storage requirements of the Imperial War Museum is dated 10 May 1944.[80] It put the off-site storage space required by the museum at 60,000 square feet, with 30,000 being at ground level.[81] Further documentation also reveal that other essential requirements of the space included cover, lighting, unfluctuating environmental conditions, minimum dimensions of 11 feet and 6 inches high and of 15 feet and 9 inches wide, and accessibility to large *matériel* such as aircraft, tanks and guns. It was also desired that the site be in one-and-a-half hours travel from the museum, reachable by public transport, fireproof and as burglar-proof as possible.[82] Accordingly, not any backyard or warehouse would be considered acceptable. With these requirements established, a long and challenging search began.

Over the next two years and three months, various possibilities were floated. These ranged considerably in venue type, their distance from the Imperial War Museum, and the overall potentiality and suitability of the site. The first possibility, arising during May 1945, was a proposed development off Brook Drive, located just behind the museum.[83] The second possibility, arising three months later, was Embassy Filling Station on Ashburnham Road in Chelsea.[84] After that fell through, Bradley started becoming increasingly concerned about the situation, describing the need as 'really desperate and urgent'.[85] This is because the service departments, brought on by their own spatial requirements, began threatening to discard the material which they had earmarked for the Imperial War Museum. It certainly increased the imperative of the search. A further three possibilities then arose in April 1946. These were Matching Aerodrome, Essex; Winkfield Aerodrome, Berkshire; and the Royal Ordinance Factory, Herefordshire.[86] But none of them amounted to anything.

During June 1946, hope was kindled with the possibility of acquiring 22,600 square feet at a Royal Air Force Maintenance Unit in Kidbrooke, east London.[87] Although far from the 60,000 square feet set out on the original specification document drawn up by the Imperial War Museum, Bradley reported to the Board of Trustees following his visit there that the site appeared 'most satisfactory'.[88] It even had ready workers which could be employed there by the museum.[89] In a cruel twist of fate however, despite positive discussions and negotiations, an executive decision by the Board of Trade saw the space reallocated to the Ministry of Labour and National Service at the eleventh hour without consulting the museum.[90] Bradley forcefully protested against this.[91] In response, the Imperial War Museum was offered storage at the Stevens Sandpits facility at Crayford, which exceeded the size of the Kidbrooke site. Fortuitously, it met all the requirements that the museum had originally laid down. It also required only moderate structural intervention to become useable.[92] The Trustees of the Imperial War Museum therefore signed the contract.[93] What was not known at signing, however, is that the work towards the necessary intervention would be delayed until September 1948, well over a year away.[94] As such, despite now possessing the necessary space on paper, the museum had to endure further wait before being able to use it. Until this time came about, all the staff could do was hope that the service departments would continue retaining the items which they had earmarked.

Any fallout from this further delay which the Imperial War Museum experienced while awaiting the Crayford site's development is a topic for another study. Suffice to say however, the delay did not unduly derail its overall reinvention. Later documentation in the museum archive, reflections in the unpublished war history of the institution, and indeed the presence today of various key Second World War era accessioned objects in the collection, provide evidence which shows that despite contrary threats, the service departments continued keeping hold of collection items until they could be accepted. It is fortunate this was the case. Aside from documenting their own histories, by continuing to retain material, the service departments secured the long-term continuance of the Imperial War Museum and, in the process, also facilitated a post-Second World War public understanding of the conflict.

The Crisis of Irrelevancy as Deep-Revolutionary Crisis

Sources of Disruption

The current chapter is the third of three chapters that together have explored the crisis-conducive situation which confronted the Imperial War Museum in the Second World War era, deriving from perceptions of post-war cultural irrelevancy and of the mitigating response by the museum to the problem. The crisis typology which this crisis-conducive situation embodied was introduced in Chap. 6 as a deep-revolutionary crisis. These are crises that impact museums non-physically. That could be by damaging the perceived relevance of an impacted museum to the general public, or it could be by damaging the museum's reputation. In either case, the result is diminishing legitimacy, which makes the museum appear decreasingly right and justifiable, wherever they operate. Having established the premise of the deep-revolutionary crisis in Chap. 6, demonstrated how the perception of post-war cultural irrelevancy influenced the course of the Imperial War Museum in Chaps. 6 and 7, and exemplified how the museum managed this crisis-conducive situation in Chaps. 6, 7 and 8, the current chapter now discusses how both the museum's perception of cultural irrelevancy and response to it illustrates the deep-revolutionary crisis typology.

Sources of disruption arising from a deep-revolutionary crisis include anything that causes an incoherence between an impacted museum and the society it inhabits, such as failure of the museum to meet the wants, needs and interests of the constituents they serve. The threat is omnipresent as social systems are constantly shifting, although the rapidity at which changes arise depends on the social system's stability. Museums which keep pace with the wants, needs and interests of society can avoid deep-revolutionary crisis. Those which fall behind become vulnerable. According to Montgomery Van Wart, stable societies undergo major value shifts over two generations, around 50 years.[95] Yet, in periods of instability, such as war and armed conflict, value shifts become more likely and unpredictable, with established institutions buckling and breaking under the strain of new social orders created out of the chaos of societal turbulence.[96] The two world wars exemplify this.[97] For example, in the United Kingdom, the 35-year period following the onset of the First World War in 1914 saw unfold two major societal shifts: the establishment of universal suffrage following 1918 and the modern welfare state, including a national health service, from 1945.[98]

The deep-revolutionary crisis-conducive source of disruption confronted by the Imperial War Museum over the Second World War era was the threat that the public would cease to recognise its founding concept or retain interest in its subject remit after the conflict. Neither represented the contemporary context from September 1939 onwards. The Imperial War Museum had been established in 1920 to present and interpret the First World War as the 'war to end all war'.[99] Through conveying this idea, it focussed on displaying the war experience of the United Kingdom, British Empire and Dominions over a very specific time period.[100] The coming of the Second World War however invalidated the museum's founding concept. It had also captured public interest away from the First World War, or so was the belief which developed at the Imperial War Museum.[101] Thus, without modification to its *raison d'être* and rational, the general public might have stopped supporting the museum.

Potential and Actual Effects

In line with a deep-revolutionary crisis, the possibility that the general public would stop supporting the Imperial War Museum potentialised significant disruption there through diminishing its acceptability. As with all organisations, museums must demonstrate public benefit to attract funding and retain support.[102] If they cannot demonstrate a public benefit, museums may find the argument for pursuing their aims and objectives, and consuming resources in the process, increasingly difficult to make. That various instances can be found where museums and other heritage attractions have closed (before the worldwide coronavirus pandemic of 2020–2022) through diminishing support, exemplifies this possibility.[103]

While the Imperial War Museum did not experience disruption through diminishing support, key individuals within and out the museum were aware of the potential, as discussed in Chap. 6. For example, the Director-General, Leslie Bradley, is previously quoted to have 'see[n] no future for the Museum if it was to remain merely a Museum of the last war but one'.[104] This also became a view at the Treasury. As one official was previously shown to believe: 'Future generations [...] would not attend in large numbers a Museum whose exhibits were limited to the last war'. Fear of this prospect focussed minds to mitigating it.[105]

Its Practical and Critical Significance

The limited practical significance of the danger of post-war cultural irrelevancy on the Imperial War Museum underscores the deep-revolutionary crisis by potentialising the reduction of institutional and organisational legitimacy. Such crises bring about disruption through causing misalignments between a museum and the social system, which in turn risks infringing the implicit contract between institution and society stipulating that museums must make relevant contributions to the latter.[106] Where social contracts are broken, diminishing social, political and economic capital follows, resulting from passive disinterest in or active questioning of the museum. This articulation holds with crisis' modernity-esque reading that something is gripped by a situation which it must rectify to become liberated from the crisis conditions.[107] The danger posed by post-war cultural irrelevancy had this very dimension for the Imperial War Museum. As discussed in Chap. 6, without intervention, the threat was perceived to potentialise conditions which would ultimately restrict or cut the flow of resources needed by the museum to continue existing and dispensing civic services.

The extensive critical (that is critiquable) significance of the threat of post war cultural irrelevancy also underscores deep-revolutionary crisis. As explained, such crises arise from a divergence between institution and society rather than some rupture in the lived environment. Hence, the museum impacted by deep-revolutionary crisis is not threatened with operational collapse, but rather has departed from the norms and/or expectations of society, with all the ramifications that causes. This necessitates a crisis critique which not only analyses the situation impacting the museum, but also circumstances beyond the museum's confines including the societal context. The nature of the museum, the nature of society, and the nature of the relationship between both are fundamental to the existence and understanding of deep-revolutionary crisis. Indeed, the Imperial War Museum did not encounter a physical rupture when confronting the threat posed by post-war cultural irrelevancy as the Second World War approached. Rather, it encountered change which the museum needed to keep pace by acknowledging and representing the new world reality.

Further illustrating the practical and critical significance of the deep-revolutionary crisis on the Imperial War Museum, an analogy can be drawn with what Marxist philosopher Antonio Gramsci conceived in political contexts as the 'organic crisis' in political contexts.[108] There are clear similarities between both scenarios. During organic crises, circumstances prompt the proletariat and other ruled classes to question the legitimacy of the authority and control previously maintained by the political ruling class and to threaten to abandon it. Thomas R. Bates interprets this as occurring because 'the people cease to believe the words of the national leaders'.[109] As a response to the collapse in trust, the ruling class 'reabsorbs the control that was slipping from its grasp' by making, so explains Gramsci, 'changes [to] men and programmes [...], with greater speed than is achieved by the subordinate class'.[110] Similar could be posited about the Imperial War Museum, which foresaw such a prospect occurring and countered the threat by reinventing. Through updating its *raison d'être* and rationale, the museum averted cynicism or disbelief in the institution's message ahead of any emerging discontinuity and debate over the merits of dissolving the museum.

Mitigating Measures

The response by the Imperial War Museum to the threat of post-war cultural irrelevancy initially involved accepting that the museum now inhabited a new and irreversible social system, which its existing *raison d'être* and rational were incompatible with. On arriving at this acceptance, the response then involved redefining the existing *raison d'être* and rational so they could become compatible once again: the core objective when mitigating deep-revolutionary crisis. Taking inspiration from the idea of revolution—where, as per Simon J. Knell, Suzanne MacLeod and Sheila Watson's framing of the situation, 'the museum sees two possible futures, one that reflects the present trajectory and one that can be obtained by reinvention'[111]—deep-revolutionary crisis requires an impacted museum to liberate itself from the problem through instigating a series of activities which draw on the transformation conceptualisation of resilience.[112] Put another way, when resolving deep-revolutionary crisis, the museum must prevent the situation from causing its collapse by reconfiguring itself such that the problem is cancelled out. The aim of deep-revolutionary crisis management therefore involves creating a new museum paradigm which accounts for the new world reality.

Exemplifying this response, on appreciating the danger, the Imperial War Museum pursued a crisis management strategy that ensured its *raison d'être* and rationale remained coherent with the wants, needs and interests of constituents. The ensuing institutional reinvention, which embodied the museum's response to the threat, encompassed two phases. The first comprised a detection phase. It was in this period when the perception of the potential misalignment between the museum and society became apparent to the museum. It commenced a response after key agents sensed and legitimised the situation. This involved determining the circumstances surrounding the danger and the resolution to the problem. Through doing so, the museum understood that what society sought following the new war was a museum which represented their experiences of the conflict. If the new war did not feature in the representation, the museum risked alienating existing audiences by appearing anachronistic. It also risked dissuading new audiences from developing. The resulting reinvention provided the capacity and capability for the museum to address the misalignment.[113] If successfully implemented, it offered a representation which people in the post-war landscape would recognise when the war ended and the museum reopened.

The second phase of the response comprised the prevention work. During this period, the staff averted the misalignment between the Imperial War Museum and society. The work involved three tasks: seeking the necessary permission and legitimacy from the appropriate stakeholders to reinvent as discussed in Chap. 6, conceiving and formalising the desired state which the museum would adopt as discussed in Chap. 7, and performing actions which saw the museum reconfigure around its vision for the desired state as discussed in the current chapter. Through reinventing, the museum assimilated the new societal context. This was not just a context wherein the framing of the First World War as the 'war to end all war' had become untenable, but also one wherein public interest with the Second World War overshadowed the First and wherein interest in the development and application of technology had increased. Consequently, the reinvention process acquired new material for the collection, which enabled it to make a representation of the new world reality. It also steered the museum towards presenting objects with increasing technological sophistication and profiling their technical attributes. By undertaking this work, the Imperial War Museum adopted a posture from which to begin satisfying the wants, needs and interests of the constituents served when the opportunity arose after the Second World War had concluded.

Conclusion

The Second World War had a profound impact on the Imperial War Museum. In addition to the operational disruption it experienced from German aerial attacks on London discussed in Chap. 5, the conflict disrupted the museum's philosophical underpinnings. Through the latter, the Imperial War Museum underwent significant ideological reframing. That is, the conflict catalysed revolution in the museum's *raison d'être* and rationale. Accordingly, the Imperial War Museum transitioned from a museum on the 'war to end all war' to a museum on the two world wars. This came about via an institutional reinvention: by establishing some new and desirable state and realigning the museum's various elements against that state.

The reinvention undertaken at the Imperial War Museum over the Second World War arose from concern about a crisis of irrelevancy if some new international conflict was to break out across Europe, especially one involving the United Kingdom. Tragically, another European and eventually world war did break out, creating a crisis-conducive situation at the museum, which clearly resembled what was conceived during Chap. 6 as deep-revolutionary crisis. Through engaging with the deep-revolutionary crisis concept over the previous three chapters, this study argues that museums are susceptible to becoming misaligned with constituent and broader public wants, needs and interests in the social system. Indeed, when they become misaligned, museums risk losing public support. This in turn risks them losing societal legitimacy. The approach to resolving such situations, as exemplified by the Imperial War Museum over the preceding case study, involves reconfiguring around local and/or national priorities as institutionally appropriate.

In contributing to this conclusion, the current chapter explores the vast, heterogeneous process of translating conceptual reinvention against a museum. It focusses on the first crucial steps towards acquiring collection material from the Second World War and obtaining the space required so the collected material could be stored. Through doing so, the chapter demonstrates that translating reinvention in the museum context is a process which will not necessarily take place smoothly. It requires an overarching strategy which accounts for the environment and enables the museum undergoing transformation to accommodate protraction in the process towards the reinvention, necessary reversals of work already undertaken, or repetition if work carried out is unintentionally undone.

The first steps undertaken by the Imperial War Museum in aligning its collection demonstrate the need for patience and reflexivity when developing a strategy. Experience collecting during the First World War showed that if the museum was to open with as complete an account and representation of the Second World War as possible, then it needed to start collecting immediately the conflict began. But the dangers posed by aerial attack and problems with limited storage space meant the museum began this work by focussing on small, replaceable records while trusting that prospective doner organisations—notably the armed forces—would earmark three-dimensional material for collection after the war. To ensure three-dimensional material could be collected after the war, the museum simultaneously took first steps towards aligning its spatial facilities in three ways: rationalising the collection, developing the museum and securing external storage. Each method had varying degrees of success, with the proposed redevelopment producing no tangible results at the time and thus prompting searches for external storage. This demonstrates the need to factor multiple methods when devising a reinvention strategy. It also demonstrates the possibilities that experimentation can bring the reinvention process. At the Imperial War Museum, developing the site whereon the museum was located potentialised multiple considerable institutional improvements which went beyond generating storage space. Yet, the risk of experimentation producing no results, as occurred with the museum's redevelopment, demonstrates the requirement for museums to be prepared to initiate further methods when the unlikeliness of the success of one becomes apparent.

Notes

1. *Emergency Powers (Defence)* Act, 1940, c. 20; Winston Churchill, Parliamentary Debates, Commons, 5th ser., vol. 364 (1940), col. 1159–1172; Franz Uhle-Wettler, 'War', in *Brassey's Encyclopedia of Military History and Biography*, ed Franklin D. Margiotta (Washington, DC: Brassey's, 1994), 1047.
2. Willie Pietersen, *Reinventing Strategy: Using Strategic Learning to Create and Sustain Breakthrough Performance* (New York, NY: John Wiley).
3. Christine M. Pearson and Ian I. Mitroff, 'From Crisis Prone to Crisis Prepared: A Framework for Crisis Management', *The Executive* 7 (1993): 52–53.

4. Richard Beckhard, 'Strategies for Large System Change', *Sloan Management Review* 16 (1975): 45.
5. Basil Hatim and Jeremy Munday, *Translation: An Advanced Resource Book* (Abingdon: Routledge, 2004).
6. Pietersen, *Reinventing Strategy*, 128.
7. Ibid., 139–141; Robert R. Janes, *Museums and the Paradox of Change: A Case Study in Urgent Adaption*, 3rd edn (Abingdon: Routledge, 2013), 274–276.
8. Bruno S. Frey, Fabian Homberg and Margit Osterloh, 'Organizational Control Systems and Pay-for-Performance in the Public Service', *Organization Studies* 34 (2013): 949–972.
9. Pietersen, *Reinventing Strategy*, 141–142.
10. Piotr Bienkowski and Hilary McGowan, *Managing Change in Museums and Galleries: A Practical Guide* (Abingdon: Routledge, 2021), 126–128.
11. Josh Linkner, *The Road to Reinvention: How to Drive Disruption and Accelerate Transformation* (San Francisco, CA: Jossey-Bass, 2014), 77.
12. Pietersen, *Reinventing Strategy*, 142–143.
13. Janes, *Paradox of Change*, 192–104.
14. Harsh Pathak, *Organisational Change* (New Delhi, NCT: Pearson Education, 2011), 30, 35–38.
15. Pietersen, *Reinventing Strategy*, 143–147.
16. Bienkowski and McGowan, *Managing Change*, 106–109.
17. Ibid., 103–105; Paul Strebel, 'Why Do Employees Resist Change?', *Harvard Business Review* 74 (1996): 86–92.
18. Pietersen, *Reinventing Strategy*, 211.
19. Ibid., 210.
20. Tracy Goss, Richard Pascale and Anthony Athos, 'The Reinvention Roller Coaster: Risking the Present for a Powerful Future', *Harvard Business Review* 71 (1993): 98.
21. Rosabeth Moss Kanter, Barry A. Stein and Todd D. Jick, *The Challenge of Organisational Change* (New York, NY: Free, 1992), 372.
22. Pietersen, *Reinventing Strategy*, 210–219.
23. John Hayes, *The Theory and Practice of Change Management*, 5th ed (London: Palgrave Macmillan, 2018), 458–461.
24. Beckhard, 'Large System Change', 45.
25. Pietersen, *Reinventing Strategy*, 141–142.
26. Ibid., 141.
27. Bruce A. Campbell and Christian Baars, 'Introduction', in *The Curation and Care of Museum Collections*, ed Bruce A. Campbell and Christian Baars (Abingdon: Routledge, 2019), xvi.
28. Janes, *Paradox of Change*, 112.

29. IWM, Museum Archive (MA), EN2/1/GUN/001/1, see documents therein.
30. IWM, MA, EN2/1/LON/001/1, Director-General to Clerk to the Council, 5 June 1940; IWM, MA, EN2/1/LON/001/1, Director-General to Clerk to the Council, 26 November 1940.
31. The National Archives (TNA), AIR 2/10188, Bradley to Ross, 24 October 1939, 2.
32. Ibid.
33. Pietersen, *Reinventing Strategy*, 210.
34. Bienkowski and McGowan, *Managing Change*, 11–13.
35. Kanter, Stein and Jick, *Organisational Change*, 372.
36. TNA, FO 371/24420, see documents therein.
37. TNA, T 162/742/3, Green to Bradley, 29 January 1940.
38. IWM, MA, EN2/1/LON/001/1, Salmon to Bradley, December 1940.
39. TNA, CO 323/1752/5, MacDonald to all Colonies, Protectorates and Mandated Territories, 28 February 1940.
40. TNA, CO 323/1752/5, 'Enclosure In Circular Despatch Dated 28 February 1940'.
41. TNA, EB 3/19, Beresford to the Trustees of the Imperial War Museum, 23 April 1940, 1.
42. TNA, EB 2/2, 'Sub-Committee on the Allocation of Relics of the Present War', 6.
43. TNA, EB 5/4, Reynolds to Stephens, 30 April 1945, 1.
44. Ibid.
45. IWM, MA, EN2/1/COB/004/4, Board of Trustees, 8 May 1945, 2.
46. TNA, EB 5/4, Bradley to Reynolds, 12 May 1945, 1.
47. IWM, MA, EN2/1/GOV/085/1, 'Collection of Material Relating to the Present War for Exhibition in the Army Galleries and for Record in the Reference Departments of the Imperial War Museum', 1.
48. Ibid.
49. Pietersen, *Reinventing Strategy*, 141–142.
50. Ibid., 142.
51. Sue Malvern, 'War, Memory and Museums: Art and Artefact in the Imperial War Museum', *History Workshop Journal* 49 (2000): 188.
52. TNA, EB 5/4, 'Allocation of Relics of the Late War', by Reynolds, 13 March 1946; Thad Parsons III, 'The Science Museum and the Second World War', in *Science for the Nation: Perspectives on the History of the Science Museum*, ed Peter J. T. Morris (Basingstoke: Palgrave Macmillan, 2010), 84.
53. Bienkowski and McGowan, *Managing Change*, 11–13.
54. TNA, EB 5/4, Disposals Board, 17 December 1945, 4; Peter Elliott, '"Flight without feathers is not easy": John Tanner and the Development

of the Royal Air Forms Museum', in *Museums, Modernity and Conflict: Museums and Collections in and of War Since the Nineteenth Century*, ed Kate Hill (Abingdon: Routledge, 2021), 191–211.

55. TNA, EB 2/2, 'Sub-Committee on the Allocation of Relics of the Present War', 2.
56. TNA, EB 5/4, Disposals Board, 17 December 1945, 3.
57. Ibid., p. 2.
58. Central Statistical Office, *Statistical Digest of the War*, History of the Second World War (London: HMSO, 1951), 9.
59. TNA, EB 5/4, Disposals Board, 17 December 1945, 4.
60. Ibid., 4–5.
61. Ibid., 5.
62. IWM, MA, EN1/1/REP/032, 'War History of the Imperial War Museum: Part II, January 1944 to the Reopening of the Museum on 27th November 1946', 23.
63. IWM, MA, EN2/1/DIS/002/2, unidentifiable individual [titled as Local Director] to Bradley, 6 July 1940; IWM, MA, EN2/1/DIS/002/2, unidentifiable individual [titled as Local Director] to Bradley, 11 March 1941; IWM, MA, EN2/1/DIS/002/2, unidentifiable individual [titled as Local Director] to Bradley, 14 May 1941; IWM, MA, EN2/1/DIS/002/2, Williams to Bradley, 20 October 1942.
64. IWM, MA, EN1/1/REP/032, 'War History of the Imperial War Museum, 1933–1943', 29–30.
65. Freda Matassa, *Valuing Your Collection: A Practical Guide for Museums, Libraries and Archives* (London: Facet, 2017), 113–114.
66. IWM, MA, EN1/1/REP/032, 'War History […], 1933–1943', 30.
67. Timothy Ambrose and Crispin Paine, *Museum Basics: The International Handbook*, 4th edn (Routledge: Abingdon, 2018), 196.
68. IWM, MA, EN1/1/REP/032, 'War History […], 1933–1943', 30; Peter Thorsheim, *Waste into Weapons: Recycling in Britain during the Second World War* (New York, NY: Cambridge University Press, 2015), 61.
69. IWM, MA, EN2/1/GOV/103, Bradley to de Norman, 10 February 1943; IWM, MA, EN2/1/GOV/103, Bradley to de Norman, 22 February 1943; IWM, MA, EN2/1/GOV/103, de Norman to Bradley, 21 April 1943.
70. IWM, MA, EN2/1/COB/002/6, Board of Trustees, 1 November 1943, 2.
71. Ibid.
72. IWM, MA, EN2/1/COB/003/1, Board of Trustees, 3 January 1944, 2.

73. IWM, MA, EN2/1/MUS/016, 'Note on the Proposed Reconstruction Scheme', by Blaikley; 10 January 1944; IWM, MA, EN2/1/COB/003/2, Board of Trustees, 6 March 1944, 2–3.
74. IWM, MA, EN2/1/GOV/103, Portal to Chatfield, 10 November 1943, 1; IWM, MA, EN2/1/COB/003/3, Board of Trustees, 1 May 1944, 1.
75. IWM, MA, EN2/1/GOV/103, Bull to Bradley, 1 July 1943; IWM, MA, EN2/1/GOV/103, Batch to Bradley, 2 July 1943; IWM, MA, EN2/1/COB/003/1, Board of Trustees, 3 January 1944; IWM, MA, EN2/1/GOV/103, Portal to Chatfield, 10 November 1943; IWM, MA, EN2/1/GOV/103, 'Imperial War Museum', by Bull, 14 January 1944.
76. IWM, MA, EN2/1/ACC/002, Bradley to Secretary, 10 May 1944.
77. Pietersen, *Reinventing Strategy*, 210–211.
78. Hayes, *Change Management*, 458–459.
79. Thomas Hall and Wally Bock, *Ruthless Focus: How to Use Key Core Strategies to Grow Your Business* (Indianapolis, IN: Dog Ear, 2010), 82.
80. IWM, MA, EN2/1/ACC/002, Director to Secretary, 10 May 1944.
81. Ibid.
82. IWM, MA, EN2/1/ACC/002, Director to the Secretary, 7 March 1946; IWM, MA, EN2/1/ACC/002, anon. to Proctor, 8 March 1946; IWM, MA, EN2/1/ACC/002, 'Imperial War Museum'.
83. IWM, MA, EN2/1/ACC/002, anon. to de Normann, 11 May 1945.
84. IWM, MA, EN2/1/ACC/002, Beeston to Bradley, 31 August 1945.
85. IWM, MA, EN2/1/ACC/002, anon. to de Normann, 15 February 1946.
86. IWM, MA, EN2/1/ACC/002, Le May to Bradley, 5 April 1946; IWM, MA, EN2/1/ACC/002, Rogers to Plews, 11 April 1946; IWM, MA, EN2/1/ACC/002, Procter to Bradley, 16 April 1946.
87. IWM, MA, EN2/1/ACC/002, Brightwell to Bradley, 3 June 1946.
88. IWM, MA, EN2/1/COB/005/4, Board of Trustees, 1 July 1946, 2.
89. IWM, MA, EN2/1/ACC/002, Brightwell to Bradley, 3 June 1946; IWM, MA, EN2/1/ACC/002, 'Kidbrooke R.A.F. Maintenance Depot', 6 June 1946, 3.
90. IWM, MA, EN2/1/ACC/002, Brightwell to Bradley, 8 October 1946, 1.
91. IWM, MA, EN2/1/ACC/002, anon. to Brightwell, 7 October 1946.
92. IWM, MA, EN2/1/ACC/002, Brightwell to Bradley, 8 October 1946.
93. IWM, MA, EN2/1/ACC/002, Staplehurst to Bradley, 1 April 1947.
94. IWM, MA, EN2/1/ACC/002, anon. to Parkes, 23 September 1948.
95. Montgomery Van Wart, 'The First Step in the Reinvention Process: Assessment', *Public Administration Review* 55 (1995): 429.
96. Arthur Marwick, *War and Social Change in the Twentieth Century* (London: Macmillan, 1974), 13.

97. Arthur Marwick, ed., *Total War and Social Change* (Basingstoke: Macmillan, 1988); Arthur Marwick, Clive Emsley and Wendy Simpson, eds, *Total War and Historical Change: Europe, 1914–1955* (Maidenhead: Open University Press, 2021).
98. Cheryl Law, *Suffrage and Power: The Women's Movement, 1918–1928* (London: I. B. Tauris, 1997); David Gladstone, *The Twentieth-Century Welfare State* (Basingstoke: Macmillan, 1999), 32–51.
99. IWM, MA, EN1/1/REP/032, 'Imperial War Museum: 21st Annual Report of the Director-General to the Board of Trustees', 1; Malvern, 'War', 181.
100. Jennifer Wellington, *Exhibiting War: The Great War, Museums, and Memory in Britain, Canada, and Australia* (Cambridge: Cambridge University Press, 2017), 237–256.
101. Imperial War Museum, *A Short Guide to the Imperial War Museum*, 3rd edn ([London?]: [HMSO?], [1946?]), 2.
102. Lynda Kelly, 'Measuring the Impact of Museums on Their Communities: The Role of the 21st Century Museum', in *New Roles and Missions of Museums: INTERCOM 2006 Symposium*, ed Ying-Ying Lai (Taipei: ICOM-INTERCOM, 2006), 25–34.
103. For example, Dan Martin, 'Please? Please? Please? Meh: Beatles Museum Closes Owing to Lack of Interest', *Guardian*, 7 June 2012, https://www.theguardian.com/music/2012/jun/07/beatles-museum-hamburg (18 March 2024); Nick Clark, 'Celebrated Modernist Home Closed to Visitors by the National Trust Due To Lack of Interest', *Independent*, 30 December 2013, https://www.independent.co.uk/arts-entertainment/architecture/who-wants-to-visit-a-house-like-this-not-enough-of-us-apparently-celebrated-modernist-home-closed-to-visitors-by-the-national-trust-due-to-lack-of-interest-9030836.html (18 March 2024); Pat Borzi, 'In Town Where Pitcher's 266 Wins Were Part of Fabric, One Big Loss', *New York Times*, 16 February 2015, https://www.nytimes.com/2015/02/17/sports/baseball/bob-feller-museum-in-van-meter-iowa-to-become-city-hall.html (18 March 2024); Dave Stewart, 'Founders Hall Interpretive Centre Closes Doors', *SaltWire*, 10 May 2016, https://www.saltwire.com/prince-edward-island/news/founders-hall-interpretive-centre-closes-doors-101756/ (18 March 2024); Laura Hill, 'Sadness as Durham Light Infantry Museum Closes Its Doors for the Final Time', *Chronicle*, 31 March 2016, https://www.chroniclelive.co.uk/news/north-east-news/gallery/sadness-durham-light-infantry-museum-11121791 (18 March 2024).
104. TNA, AIR 2/10188, Bradley to Ross, 24 October 1939, 1.
105. TNA, T 162/742/3, memorandum, by an unidentifiable individual, 9 October 1939.

106. Robin T. Byerly, 'Business IN Society: The Social Contract Revisited', *Journal of Organisational Transformation and Social Change* 10 (2013): 4–20.
107. Reinhart Koselleck, 'Crisis', trans. Michaela W. Richter, *Journal of the History of Ideas* 67 (2006): 161.
108. Antonio Gramsci, *Selection from the Prison Notebooks*, ed and trans. Quintin Hoare and Geoffrey Nowell Smith (London: Lawrence and Wishart, 1971), 210.
109. Thomas R. Bates, 'Gramsci and the Theory of Hegemony', *Journal of the History of Ideas* 36 (1975): 364.
110. Gramsci, *Prison Notebooks*, 210.
111. Simon J. Knell, Suzanne Macleod and Sheila Watson, 'Introduction', in *Museum Revolutions: How Museums Change and Are Changed*, ed Simon J. Knell, Suzanne Macleod and Sheila Watson (Abingdon: Routledge, 2007), xix.
112. Brigit Maguire and Sophie Cartwright, *Assessing a Community's Capacity to Manage Change: A Resilience Approach to Social Assessment* (Canberra, ACT: Bureau of Rural Science, 2008), 4–5, https://www.agriculture.gov.au/sites/default/files/abares/documents/dewha-resilience-sa-report-final-4.pdf (18 March 2024).
113. Christer Pursiainen, *The Crisis Management Cycle* (Abingdon: Routledge, 2018), 79–81, 85–88.

CHAPTER 9

Conclusions and Reflections

Introduction

This study explores how crises can disrupt museums and the contrasting defensive and revolutionary approaches to crisis management which museums might take when resolving crises. It advances the conceptual knowledge and understanding of museum crises and is among the first exploring museological issues and ideas in the context of organisational crisis management, change and resilience. In doing so, this study exemplifies approaches which museums can follow when overcoming different crisis typologies, specifically those conceived over the previous research: the surface-defensive and deep-revolutionary crises.

Through this study, historical knowledge and understanding of the Imperial War Museum has also been expanded. It fills the final gap in the history pertaining to the museum which, hitherto, has largely circumvented the Second World War era (notwithstanding certain specific if limited considerations) and shows how that wartime experience effected the institution's post-war development. It also develops positively the legacy of the Director-General at the time, Leslie Ripley Bradley, who led the museum during the conflict.

P. W. Deans, *Crisis, Reinvention and Resilience in Museums*,
https://doi.org/10.1007/978-3-031-67806-6_9

Study Context

The first two chapters examined the conceptual and historical context of the study. Chapter 2 began by setting out issues and ideas regarding and surrounding the concepts of crisis and organisational crisis management, as well as those feeding into them, such as reinvention and resilience. The chapter finds that the term crisis describes and conveys issues and ideas about difficult and challenging situations in the social system. These are situations which disrupt important societal objects—objects which uphold some aspect of the social system, such as museums. Their effects are usually irregular and unstable, requiring strengths and non-routine actions to mitigate and ultimately resolve them. For example, in the museum context, crises can bring about physical defects compromising the superstructure of a building, or non-physical defects compromising the relationship between institution and society. Hence, wheresoever and howsoever they materialise, crises prevent things from functioning normally. As a concept, crisis helps people to understand phenomena that impact the world around them. Moreover, it helps people situate themselves within the world and navigate the environment. Declaring a crisis, therefore, also comprises a means for individuals to express discomfort about a situation which threatens important objects—again, such as museums—and to propose a resolution to the problem: ideas which become legitimised and furthered by a community of likeminded individuals. Drawing on all these points, crisis is defined herein as an uncomfortable, unpredictable, unstable and potentially dangerous situation, where an impacted systemic entity will be disrupted, perhaps inoperably and irreparably, requiring extraordinary intervention to be overcome.

While setting out the aforementioned, this chapter finds that crises are highly subjective. The manifestation of crisis depends on the crisis conscious person's perceptions about a crisis object: the object's health/functionality and criticality to the life being lived by the crisis conscious person. It finds that crises should not be treated as objective phenomena, notwithstanding that individual conceptualisations can be studied as defined, experienced phenomenon. This is because treating crises objectively risks presenting them as commonly understood and experienced entities, which might not be the case. It also risks assigning them a degree of causality for something, which, similarly, might not be commonly agreed to. Another finding is that crises manifest in crisis objects variably. The nature of the manifestation depends on the situation being faced by the crisis object. It follows that crises are not homogeneous constructs. They can arise from different contexts and manifest in different ways, meaning no one methodology for dealing with them will work universally.

Chapter 3 then proceeded to set out the case, the Imperial War Museum. Specifically, it characterised the Imperial War Museum as an institution, organisation and museum, from the conception of the museum in 1917, to the eve of the Second World War in 1939 and the point from which the study began proper. The chapter establishes that the case was a progressive institution, organisation and museum. This arose through it being subject to gradual, reflexive change in line with societal trends, but also by engaging with innovative practices which foreshadowed future organisational paradigms and museological concerns.

Again, while setting out the aforementioned, this chapter finds that during the inter-war years, the Imperial War Museum possessed a responsive mindset, which was carried into and maintained throughout the Second World War. Through doing so, it finds a record of reflexive institutional practice there which accommodated evolving environmental conditions. This framed the subsequent findings on crisis and change at the museum during the Second World War as having roots in historical precedent. Another finding of the chapter is that despite its historical framing, the case draws resemblances with contemporary museums institutionally, organisationally and museologically. This serves to show those findings from later chapters to bear contemporary resonance as well as historical significance, enabling them to be applied against present-day cases and situations, which increases their relevance to current museum practitioners and crisis managers. The findings from Chaps. 2 and 3 are then taken and applied in support of Chaps. 4 through 8.

Museums and Resilience

The next chapter examined how the Imperial War Museum resolved a crisis of inactivity during the Second World War. Specifically, it focusses on analysing the museum's recourse to organisational resilience—an organisational quality characterising various strengths which support organisational activity during adversity—in maintaining various civic services throughout the conflict. This work occurred despite the museum being forcibly closed for much of the war. The closure transpired because the Imperial War Museum was deemed too unsafe to admit the public while hostilities persisted, given its building's vulnerability to aerial attack and the lack of sufficient local air raid shelter facilities. Even during 1940 when the museum temporarily reopened, concerns resulted in limitations being placed on the objects which could be shown, where they could be shown, and how many people could see them. That it overcame these restrictions, however, demonstrates the potential of museums to think creatively and problem solve a course around whatever challenge has beset them.

Chapter 4 finds that the Imperial War Museum resolved its wartime crisis of inactivity by identifying and implementing alternative means for delivering a civic service. This resulted from possessing and drawing on organisational resilience. At the start of the Second World War, the Imperial War Museum curated a temporary exhibition using second line exhibition material for when the general public could be formally admitted. It then engaged with individuals remotely once, from September 1940, they could not. But even as the latter became necessary, the museum continued privately to admit individuals for research purposes. At the same time, it supported the war effort by loaning or restituting items from the collection so they could be used towards various initiatives, by hosting local and national organisations and, sometimes, even by helping those organisations carry out their war work.

Through the case presented in Chap. 4 (indeed, in each case chapter herein), this study finds that organisational resilience is a quality drawing from and permeating every corner and dimension of an institution. As such, the most successful resilience derives from multiple organisational attributes, requiring significant investment to be developed. Accordingly, a holistic base of organisational resilience is not something achieved from any one organisational feature or action. For example, installing and maintaining robust physical infrastructure may represent one possibility under certain circumstances, but such an undertaking will only offer benefit where physical infrastructural resilience has any bearing. With the example of the Imperial War Museum and its civic service provision, robust physical infrastructure alone was not enough to resolve the situation. Additional attributes were also necessary to facilitate all the services it provided during the Second World War.

In this study, several attributes of organisational resilience have been considered. Alongside robust physical infrastructure just discussed, a second is the requirement for appropriately qualified and creative staff and, perhaps more importantly, organisational support enabling the staff to discover, nurture and deploy their talents and creativity, especially under pressure when such skills are most needed. This has been illustrated through the temporary exhibition opened at the Imperial War Museum in 1940, which required staff who knew what collection items were available and the narrative they could tell. And the third is the requirement for a support network comprising similar, allied and complementary organisations and groups, wherein resources and opportunities can be coordinated and shared as required. This has been illustrated by the museum's position in and/or relationship with national and local government and ability to dispose itself at the service of others.

Ultimately, resilience and crisis are not mutually exclusive concepts. As Chaps. 2 and 4 theorise and 5 through 8 demonstrate, the former enables the latter: successful crisis management relies on the strength which only holistic organisational resilience can engender in museums. This is significant, because it shows that museums should constantly strive for organisational resilience before any crisis can potentially strike and disrupt. Only those which do prepare so will ensure they are in the best position to cope with crisis.

Museums and Crises

The next two chapters examined how the Imperial War Museum resolved crises of physical destruction and irrelevancy during the Second World War. Drawing on findings from Chap. 4, they find that when overcoming such situations, museums must possess sufficient relevant knowledge, skills, abilities and, where applicable, the requisite physical toughness. No more so is this apparent than when dealing with crises conforming to the crisis typology conceptualised and exemplified in Chap. 5. Termed the surface-defensive crisis, it comprises a crisis which impacts the routine operations of museums, indeed any organisation. Such crises do not primarily impact or signify their (in)coherence with the social system.

Surface-defensive crises arise from internal events such as, for example, water leaks and electrical short circuits, or external events such as subsidence, extreme weather and, as experienced by the Imperial War Museum, armed conflict or other human interventions. This means that crises conforming to the surface-defensive crisis typology should be resolved through being pushed back or defended against. No viable alternative solution realistically exists, because the crisis does not indicate anything problematic with the impacted museum as an institution. The only internal changes that the museum should contemplate are resiliency-orientated enhancements improving its physical and non-physical structure. Literature on preparing for and dealing with disasters in museums—managing events like floods, fires and earthquakes—thus address dealing with surface-defensive crises.[1]

The surface-defensive crisis typology is developed and instantiated over Chap. 5 through an exploration of the disruption to the Imperial War Museum caused by repeated German aerial attacks against London during the Second World War. In doing so, the chapter finds that the museum inhabited a severe surface-defensive crisis from these attacks. Comprising air raids and ground-launched V-weapon strikes, it became extremely vulnerable to the collateral damage they produced, occupying the firing line of neighbouring targets. Moreover, its building's old and fragile construction potentiated the risks posed by each attack. In response, the museum tried to contain and limit the damage, with varying degrees of success. Collections were evacuated while staff protected the premises. Following the attacks, the museum initiated a programme to restore its day-to-day operations and learn from the experience for when (if) a similar (or worse) situation occurred in the near future.

Through the case presented in Chap. 5, this study finds that surface-defensive crises pose extreme danger as well as disruption to museums. They can potentially overwhelm a museum, causing their total collapse. They can also consume extensive resources in resolving them. An effective way to prevent or minimise the risk of experiencing a surface-defensive crisis, therefore, is to establish thorough signal detection schedules and to undertake prompt remedial action when and where emerging crisis potentials are discovered. As Chap. 5 demonstrates via Christine M. Pearson and Ian I. Mitroff's five-stage framework for crisis management, signal detection, followed by mitigation work where prevention has proved impossible, occupy the most important tasks in crisis management. Oftentimes, the success and failure of crisis management depends on rigorous and appropriate preparation measures that have been implemented by the museum.

All museums are vulnerable to surface-defensive crises. Although the example seen impacting the Imperial War Museum during the Second World War was irregular and dramatic (for the conceptual context), any incident which disrupts everyday operations can be significantly debilitating. And yet, while no crisis could ever be framed as desirable or fortunate, it could be reassuring from a long-term sustainability perspective (notwithstanding the problem at hand), that such crises do not signify any particular incoherence between an impacted museum and the inhabited social system. Assuming no additional, underlying issues exist, nothing about surface-defensive crises imply decreasing cultural relevancy. Also reassuring is the straightforwardness of the crisis resolution: find the cause and neutralise the problem, or strive to weather the storm where neutralisation cannot be accomplished.

Not all crises exhibit such dramatic and potentially violent features as a surface-defensive crisis, however. This study finds that some crises are more subtle and chronic, building up over protracted periods, potentially beyond people's consciousness, until they can no longer be ignored. Examples include the kind conceptualised and termed in Chap. 6 as a deep-revolutionary crisis, which arise when impacted museums experience diminishing societal support through sustaining increased and increasing cultural irrelevancy.

Deep-revolutionary crises arise from misalignment between the *raison d'être* and rationale of a museum—their mission, values and vision—and the wants, needs and interests of the communities served. They do not derive from events *per se*, but rather from shifts in the values which users and the wider public prioritise. As such, these crises are a constant threat, because societies are constantly changing. The necessary strategy for managing deep-revolutionary crisis can be traumatic, involving adjusting an impacted museum to restore its alignment with the social system. This involves casting off an old paradigm and embracing a new one, whatever the cost in time and resources. It is essential therefore that an impacted museum makes all the necessary changes to resolve the problem, otherwise the realignment may not go far enough. Save for an unlikely reversal of the societal value shift which caused the incoherence in the first place, the problem will not go away and will likely only intensify.

Chapter 6 develops and instantiates the deep-revolutionary crisis typology by exploring the perceived threat of cultural irrelevancy that confronted the Imperial War Museum following the Second World War. It also explores the institution's approach to preventing any such disruption from arising. In doing so, the chapter finds that the museum sensed a potential deep-revolutionary crisis and initiated plans for countering it, the execution of which is investigated during Chaps. 7 and 8. Over 1917–1939, the Imperial War Museum had been framed as a museum on the 'war to end all war', consistent with the widely subscribed to pre-1939 belief about the First World War. The Second World War however disproved this myth, jeopardising the museum's future. The museum therefore updated its *raison d'être* and rationale to represent the change across its collection.

Through the case presented in Chap. 6 (and 7 and 8), this study finds that deep-revolutionary crises, as with surface-defensive crises, can be damaging to museums. But unlike surface-defensive crises, deep-revolutionary crises have the added sting of being, at first, perceptibly more difficult, which can compromise their resolution. This is because they usually do not exhibit any clear, obvious crisis symptoms. Moreover, by the time their symptoms do materialise, perhaps on spreadsheets or in exhibition spaces themselves, the root problem may have become too ingrained to be resolved. It is essential, therefore, that deep-revolutionary crises are actively sought out before they can manifest. Accordingly, as discussed in Chap. 5, a museum should establish thorough, targeted signal detection schedules for comparing and assessing its *raison d'être* and rationale against the wants, needs and interests of core users and the wider public, and undertake prompt value adjustments where these are necessary. Fortunately, despite this added sting, museums can maintain far greater management over detected deep-revolutionary crises than their surface-defensive counterparts, for deep-revolutionary crises begin and end with the museum as an institution. Put another way, because deep-revolutionary crises arise through a museum's inertia and are resolved through action undertaken by that museum, the nature of their resolution is entirely museum dependent. The earlier they can be detected, the more manageable they become. This is because the smaller the necessary value adjustment, the less traumatic the experience.

As with surface-defensive crises, all museums are vulnerable to a deep-revolutionary crisis. It is, after all, an inevitability that societies will undergo some form of value shift. For the Imperial War Museum, the deep-revolutionary crisis transpired following a sudden, dramatic and significant event which radically altered perceptions about the 'war to end all war' myth on which the museum had been founded. But societal value shifts do not always transpire under such drastic, unstable circumstances. They can, and usually will, occur during far less precarious situations, where the shift occurs gradually over multiple generations. In this latter instance, deep-revolutionary crises transpire when museums have accrued a certain level of incoherence with the social system which can no longer be borne. Again, while no crisis could ever be framed as desirable or fortunate, it could be reassuring from an operational perspective (in contrast to surface-defensive crises above) that such crises do not signify any practical issues with the impacted museum. Indeed, they pose little to no risk of causing physical disruption. Also reassuring is the control that a quickly reactive museum can have over their outcome, even if the change requires time, resources and complex processes. This means that once an impacted museum has addressed the cause of the crisis, it retains the support and legitimacy to continue serving the public. Even so, what the deep-revolutionary crisis lacks in physical toll and limited control is more than made up for in their insidiousness and potential for causing total organisational collapse without dislodging a single beam or brick. Arguably, the greatest threat to any museum is the threat of being abandoned by society for other recreational offerings. Museums must therefore guard against this situation or risk falling into a predicament which could be avoided but may not be recoverable from.

Museums and Reinvention

The final two case study chapters continue examining how the Imperial War Museum resolved a deep-revolutionary crisis of irrelevancy during the Second World War. Specifically, they examine work carried out towards realising an institutional reinvention. As discussed above, museums impacted by deep-revolutionary crisis must resolve these situations by undertaking strategic, wholesale, organisational change. Under such circumstances, reinvention will occupy one course, if not the only practicable option going forward. For the Imperial War Museum, this involved extending its subject remit by principally initiating a new collection through which the Second World War could be represented. The collecting programme included not just two-dimensional material such as documents, posters and photographs, but also three-dimensional material such as large equipment, models thereof and small items, including personal effects. In doing so, it underwent a reinvention that touched both the museum's physical and non-physical elements. Very little went unaffected during the overall reinvention process. Chapter 7 contributed to this understanding by exploring the conceptualisation work by the museum towards realising its expansion. Chapter 8 did so by exploring the museum's practical efforts.

Reinvention is the act of creating something new from something which already exists. In the organisational context, it can represent the response to dissatisfaction with a museum's *status quo*. The process of reinvention requires the recognition of some new, desired state, as well as cognisance of the first steps which should be taken in acquiring that new state. To be deemed successful, the perceived benefit must outweigh the overall final cost. A definition of cost is kept deliberately broad here. It not only refers to financial cost (though critical), but also costs in time, effort and morale, amongst other relevant resources.

Through the cases presented in Chaps. 7 and 8, this study finds that making value adjustments akin to reinvention is an effective strategy for overcoming a deep-revolutionary crisis. By making value adjustments, museums can attain new states of existence and/or identities. The more extensive their adjustment, the more dramatic their change, the more revolutionary their new paradigm. Drawing on and adapting Willie Pietersen's ideas about institutional reinvention, Chaps. 7 and 8, respectively, show this to occur through a twofold process: first by undergoing a conceptual reinvention, and then by translating that conceptualisation onto a museum's organisational elements. For the Imperial War Museum, this process was so extensive that the institution emerging from the Second World War in 1945 was, conceptually, unrecognisable from the one which entered in 1939. It was also museologically different, with new exhibitions, new exhibits, and new interpretation—wiser, darker and more awe-inspiring, if not more sophisticated, than before.

Museums are capable of escaping a deep-revolutionary crisis, even if they are not comfortable with the necessary process for resolving the problem. As with resilience, the fact that long-standing museums still exist derives from them making repeated changes in form and function over individual institutional existences. Owing to the dramatic nature of the deep-revolutionary crisis experienced by the Imperial War Museum, the reinvention it underwent was sudden and intense—more like a revolution than a value adjustment. But, as with value shifts, not all reinventions will transpire so dramatically. Conceivably, most will transpire from multiple incremental changes, which over time mount to comprise a reinvention.

It could be argued that the dramatic nature of the reinvention which the Imperial War Museum underwent because of the Second World War masks the core underlying principle of organisational reinvention generally. This principal is that reinvention is not an unusual, one-time process, but an incessant process which cannot be resisted without eventually incurring severe consequence. As societies perpetually develop, the risk of the deep revolutionary crisis looms ever large over museums, especially over those which do not frequently make corresponding value adjustments. It follows that the reinvention process will remain an ever-lasting prospect for museums. Consequently, museums which want to avoid reinvention at the scale experienced by the Imperial War Museum during the Second World War should, where possible, ensure they align their value adjustments with society's value shifts, rather than hold out until the ensuing misalignment becomes unsustainable. While change is inevitable for museums, the nature of the change depends on how it is received.

Final Remarks

The case of the Imperial War Museum during the Second World War era reveals that crisis situations can impact museums in two main ways. One is through disrupting a museum's day-to-day operations. This derives from some external or internal force interfering with routine processes, procedures and accessibility and/or by causing unsustainable damage to infrastructure. Another way is through causing a schism between an institution and the stakeholders and constituents in the society wherein it operates and serves. This derives from institutional and/or societal development that moves the priorities of the museum and/or the society away from each other, causing an incoherence between the two. But the case also shows that museums are extremely responsive and capable organisations. With the right leadership, resources and support, they can overcome diverse challenges, as demonstrated by the examples presented herein.

From this study, the following four essential points can be made about the interplay between museums and crisis. Firstly, museums must ensure they possess organisational qualities that engender a holistic, base level of resilience to adversity. Without sufficient resilient attributes, museums may struggle in overcoming crises, as they may not possess the requisite organisational strength, skills or knowledge for dealing with the adversity. Secondly, museums must understand whether the uncomfortable situation they are experiencing possess crisis potentials: whether it is uncomfortable, unpredictable, unstable and potentially dangerous; disrupts the museum, perhaps inoperably and irreparably; and requires extraordinary intervention to be overcome. If the situation does indeed possess crisis potentials, then, thirdly, museums must identify whether the crisis exhibits defensive or revolutionary characteristics: whether it embodies a surface-defensive or deep-revolutionary crisis typology. Ultimately, the nature of the crisis dictates the nature of the museum's response. A surface-defensive crisis requires museums to push back against the crisis' effects. This is towards preserving the *status quo*, as the crisis does not signify anything amiss with the museum in its social, political, cultural and economic milieu. A deep-revolutionary crisis requires museums to embrace, or at least accept, the effects caused by the crisis. This is towards assuming some new paradigm that realigns the museum's *raison d'être* and rationale with the wants and needs of the society served. Unlike a surface-defensive crisis, a deep-revolutionary crisis indicates a broadening disconnect between the museum and the expectations surrounding it from stakeholders and constituents. And fourthly, museums must undertake appropriate responses. For a surface-defensive crisis, this involves neutralising the threat or containing it and limiting the effects. For a deep-revolutionary crisis, this involves instigating a programme of change to recohere the museum with the social system.

To summarise, museums are important institutional organisations. They are, however, vulnerable to crises, something which will likely, if not inevitably, be encountered during a museum's existence. Yet, with careful management, museums may resolve their crises. Depending on a sufficient development of physical and non-physical resilience to adversity, and timely, accurate detection and sensing of crisis signals, museums can implement crisis management strategies that address the effects posed by different crises. This could involve, as appropriate, strategies which preserve the museum's *status quo*, or update their *raison d'être* and rationale for some new societal context. Inertia or delay in the crisis management cycle risks consigning an impacted museum to history, for crises of any nature are unpredictable, and cannot be counted on to resolve themselves or to progress at a slow pace. Accordingly, museums must be as crisis-ready, as much of the time, as possible.

Through the decisions and actions of the institution's leadership and staff, the Imperial War Museum was not confined to history during or soon after the Second World War. Indeed, it has continually expanded its subject remit and adapted the way this subject is presented—a practice originating in the Second World War. Today, IWM presents the war history of the United Kingdom, former British Empire and the Commonwealth from 1914 to the present. Through doing so, the museum has remained an important, vibrant and much valued institution representing state, nation, community and individual. James Taylor, as Head of Research and Information at the Imperial War Museum, once wrote that 'The Museum's continued and increasing popularity—especially when much lighter museums and heritage sites are on offer elsewhere in the capital and beyond—has always relied upon the organisation's ability and willingness to change'.[2] It is essential this remains so. No longer confined to representing the two world wars, any future courses planned by IWM must be plotted with an eye on prevailing attitudes and interests regarding the history and concepts of war and armed conflict, and scope for adjustment, should the adopted course lose relevance.

Notes

1. Graham Matthews, Yvonne Smith and Gemma Knowles, *Disaster Management in Archives, Libraries and Museums* (Aldershot: Ashgate, 2009); G. E. Gorman and Sydney J. Shep, eds, *Preservation Management for Libraries, Archives and Museums* (London: Facet, 2006); Emma Dadson, *Emergency Planning and Response for Libraries, Archives and Museums* (London: Facet, 2012).
2. James Taylor, 'Interpreting the Second World War', in *The Power of Objects: Museums and World War II*, ed Esben Kjeldbæk (Edinburgh: MuseumsEtc, 2009), 53.

Bibliography

Archival Documentation

Museum Archive of IWM

EN1/1/COB/049 — 'Standing Committee and Board of Trustees: Agenda, Minutes, Correspondence, 1939'

/2

/4

EN1/1/MINI/002 — 'Ministries: Ministry of Labour'.

EN1/1/MUS/024 — 'Imperial War Museum: Correspondence, Visitors, 1936–1940'.

/10

EN1/1/MUS/033 — 'Imperial War Museum: Museum Governance, Treasury Correspondence, 1927–1953'.

EN1/1/MUS/034 — 'Imperial War Museum: Museum Governance, Future War Relics'.

EN1/1/MUSG/011 — 'Museums: London Museum'.

EN1/1/MUSG/013 — 'Museums: National Maritime Museum'.

EN1/1/MUSG/021 — 'Museums: Overseas, Australian War Memorial'.

EN1/1/REP/003 — 'Annual Reports: 1919–1920'.

EN1/1/REP/004 — 'Annual Reports: 1920–1921'.

EN1/1/REP/005 — 'Annual Reports: 1921–1922'.

P. W. Deans, *Crisis, Reinvention and Resilience in Museums*,
https://doi.org/10.1007/978-3-031-67806-6

EN1/1/REP/006 — 'Annual Reports: 1922–1923'.
EN1/1/REP/007 — 'Annual Reports: 1923–1924'.
EN1/1/REP/008 — 'Annual Reports: 1924–1925'.
EN1/1/REP/009 — 'Annual Reports: 1925–1926'.
EN1/1/REP/010 — 'Annual Reports: 1926–1927'.
EN1/1/REP/011 — 'Annual Reports: 1927–1928'.
EN1/1/REP/012 — 'Annual Reports: 1928–1929'.
EN1/1/REP/013 — 'Annual Reports: 1929–1930'.
EN1/1/REP/015 — 'Annual Reports: 1930–1931'.
EN1/1/REP/032 — 'Annual Report Sets (4)'.
EN1/1/STA/004 — 'Staff: Duty Officers and Instructions'.
EN2/1/ACC/002 — 'Accommodation: Crayford'.
EN2/1/ACC/004 — 'Accommodation: Storage Wartime'.
/7
EN2/1/COB/001 — 'Standing Committee and Board of Trustees: Meetings and Draft Minutes, January 1940–July 1942'.
/1
/3
EN2/1/COB/002 — 'Standing Committee and Board of Trustees: Draft Minutes January–November 1943'.
/6
EN2/1/COB/003 — 'Standing Committee and Board of Trustees: Draft Minutes Feb–Nov 1944'.
/1
/2
/3
EN2/1/COB/004 — 'Standing Committee and Board of Trustees: Draft Minutes Jan–Oct 1945'.
/4
EN2/1/COB/005 — 'Standing Committee and Board of Trustees: Draft Minutes, January–November 1946'.
/2
/4
EN2/1/COB/009 — 'Board of Trustees: Draft Minutes Feb–Dec 1950'.
/2
EN2/1/COB/039 — 'Standing Committee and Board of Trustees: Charles ffoulkes 1939–1949'.
EN2/1/COB/040 — 'Standing Committee and Board of Trustees: Sir John Forsdyke'.
EN2/1/COB/049 — 'Standing Committee and Board of Trustees: Colonel E. E. B. Mackintosh'.

EN2/1/COB/051 — 'Standing Commission & Board of Trustees: Lord Melchett'.
/1
EN2/1/CON/004 — 'Conferences: Conference of the Directors of National Museums and Galleries – Air Raid Precautions (Preparations for), 1933–1934'.
EN2/1/DIS/001 — 'Disposals: General'.
EN2/1/DIS/002 — 'Disposals'.
/2
EN2/1/ENQ/001 — 'Enquiries: General 1'.
EN2/1/ENQ/024 — 'Enquiries: General Correspondence, 1940s'.
EN2/1/GOV/011 — 'Government Departments: Admiralty– Loan of Guns etc. to Chatham'.
EN2/1/GOV/058 — 'Government Departments: Ministry of Fuel and Power'.
/1
EN2/1/GOV/085 — 'Government Departments: War Office– Correspondence 1939–1960'.
/1
EN2/1/GOV/097 — 'Government Departments: Ministry of Works – General 1940–1957'.
EN2/1/GOV/103 — 'Government Departments: Ministry of Works – Correspondence re Extension of the Museum Building'.
EN2/1/GOV/111 — 'Government Departments: Ministry of Works – Storage of National Art Treasures'.
EN2/1/GUN/001 — 'Guns'.
/1
EN2/1/LON/001 — 'London County Council: 1940–1960'.
/1
EN2/1/MUS/001 — 'Imperial War Museum: War History 1933–1946'.
/1
EN2/1/MUS/002 — 'Imperial War Museum: Air Raid Precautions'.
/2
/3
/5
/16
EN2/1/MUS/003 — 'Imperial War Museum: Reopening'.
/1b
EN2/1/MUS/016 — 'Imperial War Museum: Reconstruction'.
EN2/1/STA/001 — 'Staff, General'.
EN2/1/STA/009 — 'Staff Applications'.
/4

The National Archives of the United Kingdom

AIR 2/10188 — 'Imperial War Museum: Preservation of Documents etc. Relating to the 1939–1945 War'.
CO 323/1752/5 — 'Imperial War Museum: Contributions from the Colonies'.
EB 2/1 — 'Standing Commission on Museums and Galleries: Minutes of Meetings 1–20'.
EB 2/2 — 'Standing Commission on Museums and Galleries: Minutes of Meetings 21–47'.
EB 3/17 — 'Consideration of Partial Re-opening of Certain National Museums and Galleries'.
EB 3/19 — 'Wartime Precautions'.
EB 5/4 — '58–144'.
FO 371/24420 — 'Munich Agreement - Request by Imperial War Museum'.
HO 186/2097 — 'Civil Defence Material for Imperial War Museum'.
HO 196/4 — 'Research and Experiments Department Notes'.
T 162/742/3 — 'Histories: Records of the War, 1939; General Questions Including the Utilisation of the Imperial War Museum, and Distribution of Printed Material to Outside Bodies'.
T 218/207 — 'Imperial War Museum: Museum and Tower Armouries; appointment, pay and conditions of service for Curator, Secretary, Accounting Officer and junior staff'.
WO 339/54103 — '2/Lieutenant Leslie Ripley BRADLEY The Duke of Cambridge's Own (Middlesex Regiment)'.
WORK 17/180 — 'Evacuation of art treasures to 1. Penn House, Amersham, Buckingham; 2. Colworth House, Sharnbrook, Bedford; and 3. 'Ramster', Chiddingfold, Surrey'.

Hansard Records

Conway, Martin. *Parliamentary Debates*, Lords, 5th series, volume 86 (1932).
Churchill, Winston. *Parliamentary Debates*, Commons, 5th series, volume 364 (1940).
Mond, Alfred. *Parliamentary Debates*, Commons, 5th series, volume 127 (1920).
Ormsby-Gore, William. *Parliamentary Debates*, Commons, 5th series, volume 291 (1934).
Parliamentary Debates, Commons, 5th series, volume 103 (1918).
Baldwin, Stanley. *Parliamentary Debates*, Commons, 5th series, volume 270 (1932).

Acts of Parliament

Emergency Powers (Defence) Act, 1940, chapter 20.
Imperial War Museum Act, 1920, chapter 16.
Imperial War Museum Act, 1955, chapter 14.
Museums and Galleries Act, 1992, chapter 44.
National Maritime Museum Act, 1934, chapter 43.

Printed Material, Electronic Sources and Conference Papers

Aardman. *Flight of the Stories—Our New First World War Galleries*. YouTube video. IWM, https://www.youtube.com/watch?v=-lZUTMS8uLA.

Abt, Jeffrey. *Valuing Detroit's Art Museum: A History of Fiscal Abandonment and Rescue*. Cham: Palgrave Macmillan, 2017.

Adams, Marianna and Judy Koke. 'Comprehensive Interpretive Plans: A Framework of Questions'. *Journal of Museum Education* 33 (2008): 293–300.

Adorno, Theodor W. 'Education after Auschwitz', translated Henry W. Pickford. In *Can One Live After Auschwitz? A Philosophical Reader*, edited by Rolf Tiedemann. Stanford, CA: Stanford University Press, 2003, 19–33.

Akpanama, Ntak George and Chukwuemeka Ekenta. 'Examining the Negative Impact of Abattoir Operations on Residential Property Value within Abattoir Site in Port Harcourt Metropolis'. *International Journal of Innovative Science, Engineering and Technology* 9 (2022): 26–34.

Al Luhaidan, Ibrahim and Daifallah M. Alrazeeni. 'The Role of Key Early Warning Indicators in Crisis Management: Analytical Review'. *Advanced Emergency Medicine* 8 (2019): 9–16.

Aldgate, Anthony and Jeffrey Richards. *Britain Can Take It: The British Cinema in the Second World War*. New edition. London: I. B. Tauris, 2007.

Ambrose, Timothy and Crispin Paine. *Museum Basics: The International Handbook*. 4th edition. Routledge: Abingdon, 2018.

Ambrose, Timothy. 'The Purpose of a Forward Plan: Solving Problems; Making the Best Use of Resources; Getting Value for Money'. In *Forward Planning: A Handbook for Business, Corporate and Development Planning for Museums and Galleries*, edited by Timothy Ambrose and Sue Runyard. London: Routledge, 1991, 2–5.

Anderson, Erika. 'Learning to Learn'. *Harvard Business Review* 94 (2016): 98–101.

Anderson, Gail. 'A Framework: Reinventing the Museum'. In *Reinventing the Museum: The Evolving Conversation on the Paradigm Shift*, edited by Gail Anderson. Lanham, MD: AltaMira, 2012, 1–9.

Ansell, Christopher and Martin Bartenberger. *Pragmatism and Political Crisis Management: Principle and Practical Rationality During the Financial Crisis.* Cheltenham: Edward Elgar, 2019.

Apsel, Joyce. *Introducing Peace Museums.* Abingdon: Routledge, 2016.

Argyris, Chris and Donals A. Schön. *Organizational Learning: A Theory of Action Perspective.* Reading, MA: Addison-Wesley, 1978.

Ashford, Doug. 'The Exhibition as an Artistic Medium'. *Art Journal* 57 (1998): 28–37.

Assmann, Aleida. 'Canon and Archive'. In *The Collective Memory Reader*, edited by Jeffrey K. Olick, Vered Vinitzky-Seroussi and Daniel Levy. New York, NY: Oxford University Press, 2011, 334–337.

Atkin, Nicholas; Michael Biddiss and Frank Tallett. *The Wiley-Blackwell Dictionary of Modern European History Since 1789.* Chichester: John Wiley, 2011.

Aulich, Jim and John Hewitt. *Seduction or Instruction? First World War Posters in Britain and Europe.* Manchester: Manchester University Press, 2007.

Babecký, Jan; Tomáš Havránek; Jakub Matějů; Marek Rusnák; Kateřina Šmídková; and Bořek Vašíček. 'Leading Indicators of Crisis Incidence: Evidence from Developed Countries'. *European Central Bank Working Paper Series* 1486 (2012): 1–28.

Barlow, David H. *Anxiety and Its Disorders: The Nature and Treatment of Anxiety and Panic.* 2nd edition. New York, NY: Guilford, 2004.

Bates, Thomas R. 'Gramsci and the Theory of Hegemony'. *Journal of the History of Ideas* 36 (1975): 351–366.

Beckhard, Richard. 'Strategies for Large System Change'. *Sloan Management Review* 16 (1975): 43–55.

Bell, Amy. 'Landscapes of Fear: Wartime London, 1939–1945'. *Journal of British Studies* 48 (2009): 153–75.

Bennett, Tony. *The Birth of the Museum: History, Theory, Politics.* London: Routledge, 1995.

Benny, Daniel J. *Cultural Property Security: Protecting Museums, Historic Sites, Archives, and Libraries.* Boca Raton, FL: CRC, 2013.

Berger, Peter and Thomas Luckmann. *The Construction of Reality: A Treatise in the Sociology of Knowledge.* London: Penguin, 1991.

Berger, Stefan; Anna Cento Bull; Cristian Cercel; David Clarke; Nina Parish; Eleanor Rowley; and Zofa Wóycicka. 'Memory Cultures of War in European War Museums'. In Agonistic Memory and the Legacy of 20th Century Wars in Europe, edited by Stefan Berger and Wulf Kansteiner. Cham: Palgrave Macmillan, 2021, 69–114.

Bernstein, Jonathan. *Manager's Guide to Crisis Management.* New York, NY: McGraw-Hill, 2011.

Bevan, Robert. *The Destruction of Memory: Architecture at War.* 2nd edition. London: Reaktion, 2016.

Bhamra, Ran S.; Kevin Burnard and Samir S. Dani. 'Resilience: A Concept, a Literature Review and Future Directions'. In *Organisational Resilience: Concepts, Integration and Practice*, edited by Ran Bhamra. Boca Raton, FL: CRC, 2015, 3–30.

Bienkowski, Piotr and Hilary McGowan. *Managing Change in Museums and Galleries: A Practical Guide*. Abingdon: Routledge, 2021.

Bienkowski, Piotr. 'Authority and the Power of Place: Exploring the Legitimacy of Authorized Voices in the Restitution Discourse'. In *Museums and Restitution*, edited by Louise Tythacott and Kostas Arvanitis. Farnham: Ashgate, 2014, 37–52.

Black, Graham. 'Introduction: The Challenge of Change'. In *Museums and the Challenge of Change: Old Institutions in a New World*, edited by Graham Black. Abingdon: Routledge, 2021, 1–17.

Blaikley, Ernest. 'Air Raids and the Imperial War Museum'. *Museums Journal* 40 (1941): 297–298.

Blaikley, Ernest. 'Reopening of the Imperial War Museum'. *Museums Journal* 40 (1940): 8–9.

Blank, Laurie R. and Gregory P. Noone. *International Law and Armed Conflict: Fundamental Principles and Contemporary Challenges in the Law of War*. New York, NY: Wolters Kluwer Law and Business, 2013.

Bogle, Elizabeth. *Museum Exhibition: Planning and Design*. Lanham, MD: AltaMira, 2013.

Boin, Arjen; Paul 't Hart; Eric Stern and Bengt Sundelius. *The Politics of Crisis Management: Public Leadership Under Pressure*. 2nd edition. Cambridge: Cambridge University Press, 2017.

Bonnafous-Boucher, Maria and Jacob Dahl Rendtorff. *Stakeholder Theory: A Model for Strategic Management*. [Cham?]: Springer, 2016.

Borghini, Andrea. 'A Reassessment of the Category of Crisis: Some Reflections from the Social Sciences'. In *Exploring the Crisis: Theoretical Perspectives and Empirical Investigations*, edited by Andrea Borghini and Enrico Campo. Pisa: Pisa University Press, 2015, 326–343.

Borzi, Pat. 'In Town Where Pitcher's 266 Wins Were Part of Fabric, One Big Loss'. *New York Times*, 16 February 2015. https://www.nytimes.com/2015/02/17/sports/baseball/bob-feller-museum-in-van-meter-iowa-to-become-city-hall.html.

Bourrier, Mathilde. 'Organizing Maintenance Work at Two Nuclear Power Plants'. *Journal of Contingencies and Crisis Management* 4 (1996): 104–112.

Brink, Chris. *The Soul of a University: Why Excellence Is Not Enough*. Bristol: Bristol University Press, 2018.

Brit, Thomas W. and Gargi Sawhney. 'Resilience Capacity, Processes and Demonstration at the Employee, Team and Organizational Levels: A Multilevel Perspective'. In *Research Handbook on Organizational Resilience*, edited by

Edward H. Powley, Brianna Barker Caza and Arran Caza. Cheltenham: Edward Elgar, 2020, 10–24.

Brown, Charlotte; Erica Seville and John Vargo. 'Measuring the Organizational Resilience of Critical Infrastructure Providers: A New Zealand Case Study'. *International Journal of Critical Infrastructure Protection* 18 (2017): 37–49.

Brown, Frederic J. *Chemical Warfare: A Study in Restraint*. Edition with new introduction. New Brunswick, NJ: Transaction, 2006.

Brumgardt, John R. 'Mitigating Staff Stress in a Natural Disaster: The Charleston Museum, Charleston, S.C.'. In *Institutional Trauma: Major Change in Museums and Its Effect on Staff*, edited by Elaine Heumann Gurian. Washington, DC: American Association of Museums, 1995, 59–74.

Buchanan, Allen. 'Institutional Legitimacy'. In *Oxford Studies in Political Philosophy*, volume 4, edited by David Sobel, Peter Vallentyne and Steven Wall. Oxford: Oxford University Press, 2018, 53–79.

Buchanan, David A. and Andrzej A. Huczynski (2017). *Organizational Behaviour*. 9th edition. Harlow: Pearson Education.

Bucheli, Marcelo and Jin Uk Kim. 'The State as a Historical Construct in Organizational Studies'. In *Organizations in Time: History, Theory, Methods*, edited by Marcelo Bucheli and R. Daniel Wadhwani. Paperback edition. Oxford: Oxford University Press, 2015, 241–262.

Budgen, David. 'Literature'. In *1914–1918-online. International Encyclopedia of the First World War*, edited by Ute Daniel, Peter Gatrell, Oliver Janz, Heather Jones, Jennifer Keene, Alan Kramer and Bill Nasson. Berlin, BE: Freie Universität Berlin, 2014, https://doi.org/10.15463/ie1418.10329/1.1, https://encyclopedia.1914-1918-online.net/pdf/1914-1918-Online-literature-2018-06-29-V1.1.pdf.

Burcaw, G. Ellis. *Introduction to Museum Work*. 3rd edition. Walnut Creek, CA: AltaMira, 1997.

Byerly, Robin T. 'Business IN Society: The Social Contract Revisited'. *Journal of Organisational Transformation and Social Change* 10 (2013): 4–20.

Byrne, John A. 'The Virtual Corporation'. *International Business Week*, 8 February 1993.

Cady, Steven H.; Robert 'Jake' Jacobs; Ron Koller and John Spalding. 'The Change Formula: Myth, Legend, or Lore?' *OD Practitioner* 46 (2014): 32–39.

Calder, Ritchie. 'Bernal at War'. In *J. D. Bernal: A Life in Science and Politics*, edited by Brenda Swann and Francis Aprahamian. London: Verso, 1999, 160–190.

Campbell, Bruce A. and Christian Baars. 'Introduction'. In *The Curation and Care of Museum Collections*, edited by Bruce A. Campbell and Christian Baars. Abingdon: Routledge, 2019, xiv–xix.

Campbell, Christie. *Target London: Under Attack from the V-weapons During WWII*. London: Abacus, 2013.

Campbell, Ross. *Crisis Control: Preventing and Managing Corporate Crises*. Sydney, NSW: Prentice Hall, 1999.

Candlin, Fiona. 'Defining Museums'. Blog. Mapping Museums Lab. https://mapping-museums.bbk.ac.uk/2017/10/30/defining-museums/.

Cannon, Mark D. and Amy C. Edmondson. 'Confronting Failure: Antecedents and Consequences of Shared Beliefs About Failure in Organizational Work Groups'. *Journal of Organizational Behaviour*, 22 (2001): 161–177.

Carlisle, Countess of, Esmé. *A History of the Commission*. London: Museums and Galleries Commission, 1991.

Carmeli, Abraham and John Schaubroeck. 'Organisational Crisis-Preparedness: The Importance of Learning from Failures'. *Long Range Planning* 41 (2008): 177–196.

Cassar, George. *Lloyd George at War, 1916–1918*. London: Anthem, 2009.

Central Statistical Office. Statistical Digest of the War. History of the Second World War. London: HMSO, 1951.

Cerquetti, Mara and Eleonora Cutrini. 'Structure, People and Relationships: A Multidimensional Method to Assess Museum Resilience'. *Nonprofit and Voluntary Sector Quarterly* 52 (2023): 130–152.

Cerquetti, Mara. 'Vingt ans après. Do Museum Networks Really Work? A Focus on the Inland Areas of the Marche Region's "Seismic Crater" (Italy)'. *Il Capitale Culturale. Studies on the Value of Cultural Heritage* 19 (2019): 27–51.

Chanter, Barrie and Peter Swallow. *Building Maintenance Management*. 2nd edition. Oxford: Blackwell, 2007.

Chapman, Richard A. *Civil Service Commission, 1855–1991: A Bureau Biography*. London: Routledge, 2004.

Charman, Terry. '"A Museum of Man's Greatest Lunatic Folly": The Imperial War Museum and its Commemoration of the Great War, 1917-2008'. In *A Part of History: Aspects of the British Experience of the Western Front*. London: Continuum, 2008, 99–106.

Child, John. *Organization: Contemporary Principles and Practice*. Malden, MA: Blackwell, 2005.

Clark, Andy. 'Magic Words: How Language Augments Human Computation'. In *Language and Thought: Interdisciplinary Themes*, edited by Peter Carruthers and Jill Boucher. Cambridge: Cambridge University Press, 1998, 162–183.

Clark, Nick. 'Celebrated Modernist Home Closed to Visitors by the National Trust Due To Lack of Interest'. *Independent*, 30 December 2013. https://www.independent.co.uk/arts-entertainment/architecture/who-wants-to-visit-a-house-like-this-not-enough-of-us-apparently-celebrated-modernist-home-closed-to-visitors-by-the-national-trust-due-to-lack-of-interest-9030836.html.

Clarke, Liz. *The Essence of Change*. New York, NY: Prentice Hall, 1994.

Committee of Public Accounts. *Special Report and First, Second and Third Reports from the Committee of Public Accounts Together with the Proceedings of the Committee, Minutes of Evidence, Appendices and Index.* London: HMSO, 1957.

Condell, Diana M. 'The Imperial War Museum, 1917–1920: A Study of the Institution and Its Presentation of the First World War'. Master's thesis, no place, 1985.

Condell, Diana. 'The History and Role of the Imperial War Museum'. In *War and the Cultural Construction of Identities in Britain*, edited by Barbara Korte and Ralf Schneider. Amsterdam: Rodopi, 2002, 25–38.

Conlin, Jonathan. *The Nation's Mantlepiece: A History of the National Gallery.* London: Pallas Athene, 2006.

Cooke, Steven and Lloyd Jenkins. 'Discourses of Regeneration in Early Twentieth-Century Britain: From Bedlam to the Imperial War Museum'. *Area* 33 (2001): 382–390.

Coombs, W. Timothy. *Ongoing Crisis Communication: Planning, Managing, and Responding.* 5th edition. Thousand Oaks, CA: SAGE, 2019.

Cooper, Carry; Jill Flint-Taylor and Michael Pearn. *Building Resilience for Success: A Resource for Managers and Organisations.* Basingstoke: Palgrave Macmillan, 2013.

Cornish, Paul. '"Sacred Relics": Objects in the Imperial War Museum, 1917–1939'. In *Matters of Conflict: Material Culture, Memory and the First World War*, edited by Nicholas J. Saunders. Abingdon: Routledge, 2004, 35–50.

Cornish, Paul. 'Extremes of Collecting at the Imperial War Museum, 1917–2009: Struggles with the Large and the Ephemeral'. In *Extreme Collecting: Challenging Practices for 21st Century Museums*, edited by Graeme Were and J. C. H. King. New York, NY: Berghahn, 2012, 157–167.

Cornish, Paul. 'Introduction'. In *Curating the Great War*, edited by Paul Cornish and Nicholas J. Saunders. Abingdon: Routledge, 2022, 1–18.

Counts, Charity M. 'Spectacular Design in Museum Exhibitions'. *Curator* 52 (2009): 273–288.

Coutu, Diane L. 'How Resilience Works'. *Harvard Business Review* 80 (2002): 46–55.

Crandall, William 'Rick'; John A. Parnell and John E. Spillan. *Crisis Management: Leading in the New Strategy Landscape.* Self-published 4th edition. No place: no publisher, 2021. Note that this monograph was formally published by SAGE in 2014.

Crisis Management. Boston, MA: Harvard Business School Press, 2004.

Cundy, Alys. 'A Century of Reinvention: Display Policy and Practice at the Imperial War Museum, London, 1917–2017'. Doctoral thesis, University of Bristol, 2015.

Cundy, Alys. 'Objects of War: The Response of the Imperial War Museum, London, to the First and the Second World Wars'. *Post-Medieval Archaeology* 51 (2017): 261–273.

Cundy, Alys. 'The Imperial War Museum and the Material Culture of the First World War, 1917–2014'. In *The Edinburgh Companion to the First World War*, edited by Ann-Marie Einhaus and Katherine Isobel Baxter. Edinburgh: Edinburgh University Press, 2017, 402–418.

Cundy, Alys. 'Thresholds of Memory: Representing Function through Space and Object at the Imperial War Museum, London, 1918-2014'. *Museum History Journal* 8 (2015): 247–268.

Cundy, Alys. 'War, Memory and the Senses in the Imperial War Museum, London'. In *Modern Conflict and the Senses*, edited by Nicholas J. Saunders and Paul Cornish. Abingdon: Routledge, 2017, 361–374.

Dadson, Emma. *Emergency Planning and Response for Libraries, Archives and Museums.* London: Facet, 2012.

Danzig, Richard and Peter Szanton. *National Service: What Would It Mean?* Lexington, MA: Lexington Books, 1986.

Davies, Andrew. 'The War Years'. In *Theatre and War, 1933–1945: Performance in Extremis*, edited by Michael Balfour. New York, NY: Berghahn, 2001, 54–64.

Davis, Richard. 'The British Peace Movement in the Interwar Years'. *Revue Française de Civilisation Britannique* 22 (2017): https://doi.org/10.4000/rfcb.1415, https://journals.openedition.org/rfcb/1415.

Dean, David. *Museum Exhibition: Theory and Practice.* Paperback edition. London: Routledge, 1996.

Deans, Philip W. 'Curating the Great War at the Imperial War Museum during the Second World War'. In *Curating the Great War*, edited by Paul Cornish and Nicholas J. Saunders. Abingdon: Routledge, 2022, 86–104.

Deans, Philip W. 'Protective Measures Before the 1954 Hague Convention: The Case of the Imperial War Museum, 1933–1950'. In *Safeguarding Cultural property and the 1954 Hague Convention: All Possible Steps*, edited by Emma Cunliffe and Paul Fox. Woodbridge: Boydell, 2022, 57–76.

Deans, Philip W. and Darren Kelsey. 'National Narration and Organisational Crisis: A Discourse-Mythological Analysis of the Imperial War Museum During the Second World War'. *CADAAD Journal* 15 (2024): 86–108.

Debord, Guy. *Society of the Spectacle.* London: Rebel/Dark Star, 1977.

Decter, Joshua. 'Will There Still Be a Future When the Museum of the Future Arrives?' In *The Future of Museums*, edited by Gerald Bast, Elias G. Carayannis and David F. J. Campbell. Cham: Springer International, 2018, 15–36.

Devlin, Edward S. *Crisis Management Planning and Execution.* Boca Raton, FL: Auerbach, 2007.

Dobbin, Frank R. 'Cultural Models of Organisation: The Social Construction of Rational'. In *The Sociology of Culture: Emerging Theoretical Perspectives*, edited by Diana Crane. Cambridge, MA: Blackwell, 1994, 117–141.

Douglas, Diane L. 'Historical Analogies, Use of'. In *Encyclopedia of Crisis Management*, edited by K. Bradley Penuel, Matt Statler and Ryan Hagen. Thousand Oaks, CA: SAGE, 2013, 469–474.

Douglas, Mary. *Purity and Danger: An Analysis of Concepts of Pollution and Taboo.* New York, NY: Frederick A. Praeger, 1966.

Dowling, John and Jeffrey Pfeffer. 'Organizational Legitimacy: Social Values and Organizational Behaviour'. *Pacific Sociological Review* 18 (1975): 122–136.

Duchek, Stephanie. 'Organizational Resilience: A Capability-based Conceptualization'. *Business Research* 13 (2020): 215–246.

Duran, Jasmine N. 'Headliners: The Influence of Media Coverage in Museum Crises'. Master's thesis, The University of the Arts, 2014.

Edson, Gary and David Dean. *The Handbook for Museums.* London: Routledge, 1996.

Elieli, Rina Bar-Lev and Laurence J. Gould. 'Introduction'. In *Institutional Trauma: Major Change in Museums and its Effect on Staff*, edited by Elaine Heumann Gurian. Washington, DC: American Association of Museums, 1995, 23–30.

Elliot, Anthony. *Reinvention.* Abingdon: Routledge, 2013.

Elliott, Peter. '"Flight without feathers is not easy": John Tanner and the Development of the Royal Air Force Museum'. In *Museums, Modernity and Conflict: Museums and Collections in and of War Since the Nineteenth Century*, edited by Kate Hill. Abingdon: Routledge, 2021, 191–211.

Faber, David. *Munich: The 1938 Appeasement Crisis.* London: Pocket, 2009.

Fairley, Catherine M. 'Volunteerism, Youth and the Transition to Democracy in Malawi'. *VOSESA Focus* 1 (2006): http://www.vosesa.org.za/focus/vol1_no3/index.html?article_3.html~content.

Falk, John. 'The Visitor Museum Experience: Who Visits, Why and to What Effect?' In *Reinventing the Museum: The Evolving Conversation on the Paradigm Shift*, edited by Gail Anderson. Lanham, MD: AltaMira, 2012, 317–329.

Falode, Adewunmi J. and Moses J. Yakubu. 'A New Historiography of the Origins of the Cold War'. *Jurnal Sosial dan Humaniora* 9 (2019): 100–111.

Farrell-Banks, David. *Affect and Belonging in Political Uses of the Past.* Abingdon: Routledge, 2023.

Ferguson, Niall. *Doom: The Politics of Catastrophe.* London: Allen Lane, 2021.

Ferris, Timothy L. J.; Scott Jackson; Keith D. Willett; Victoria A. Hailey; and Eric Specking. 'System Resilience: Application Domain Perspectives'. *INCOSE International Symposium* 29 (2019): 1093–1107.

Fink, Stephen. *Crisis Management: Planning for the Inevitable.* New York, NY: American Management Association, 1986.

Fleming, David. 'The Essence of the Museum: Mission, Values, Vision'. In *Museum Practice*, edited by Conal McCarthy. The International Handbooks of Museum Studies. Chichester: Wiley Blackwell, 2015, 3–25.

Fopp, Michael A. *Managing Museums and Galleries.* London: Routledge, 1997.

Foss, Brian. *War Paint: Art, War, State and Identity in Britain, 1939–1945.* New Haven, CT: Yale University Press, 2007.

Foster, H. 'The Imperial War Museum: Its New Home at Southwark'. *Museums Journal* 36 (1936): 215–222.

Foucault, Michel. 'Of Other Spaces', translated by Jay Miskowiec. *Diacritics* 16 (1986): 22–27.

Fox, James. *British Art and the First World War, 1914–1924.* Cambridge: Cambridge University Press, 2015.

'Framed'. *Daily Mirror*, 7 February 1940.

Frandsen, Finn and Winni Johansen. 'Crisis Communication and Organizational Legitimacy'. In *Handbook of Business Legitimacy: Responsibility, Ethics and Society*, edited by Jacob Dahl Rendtorff. Cham: Springer Nature, 2020, 707–725.

Frandsen, Finn and Winni Johansen. *Organizational Crisis Communication: A Multivocal Approach.* London: SAGE, 2017.

Frankland, Noble. *History at War: The Campaigns of an Historian.* London: Giles de la Mare, 1998.

Frankland, Noble. Review of Museums and the First World War: A Social History, by Gaynor Kavanagh. *Journal of the History of Collections* 7 (1995): 126–127.

Franks, George R. 'Damage Containment'. In *Encyclopedia of Crisis Management*, edited by K. Bradley Penuel, Matt Statler and Ryan Hagen. Thousand Oaks, CA: SAGE, 2013, 226–230.

Fraser, John. 'A Discomforting Definition of Museum'. *Curator: The Museum Journal* 62 (2019): 501–504.

Frey, Bruno S.; Fabian Homberg and Margit Osterloh. 'Organizational Control Systems and Pay-for-Performance in the Public Service'. *Organization Studies* 34 (2013): 949–972.

Fulbrook, Mary. *Historical Theory.* London: Routledge, 2002.

Furusten, Staffan. *Institutional Theory and Organisational Change.* Cheltenham: Edward Elgar, 2013.

Gabriel, Yinnis. *Storytelling in Organizations: Facts, Fictions, and Fantasies.* Oxford: Oxford University Press, 2000.

Garayev, Vener. 'Crisis, Definition of'. In *Encyclopedia of Crisis Management*, edited by K. Bradley Penuel, Matt Statler and Ryan Hagen. Thousand Oaks, CA: SAGE, 2013, 186–187.

Garcia, Helio Fred. 'Effective Leadership Response to Crisis'. *Strategy and Leadership* 34 (2006): 4–10.

Geisbusch, Jan. 'For Your Eyes Only? The Magic Touch of Relics'. In *Museum Objects: Experiencing the Properties of Things*, edited by Sandra H. Dudley. Abingdon: Routledge, 2012, 202–213.

Geller, Stephanie Lessans and Lester M. Salamon (with the technical assistance of Kasey L. Mengel). 'Museums and Other Nonprofits in the Current Recession: A Story of Resilience, Innovation, and Survival'. *Journal of Museum Education* 35 (2010): 129–140.

Genoways, Hugh H. and Lynne M. Ireland. *Museum Administration: An Introduction*. Walnut Creek, CA: AltaMira, 2003.

Gibbs, N. H. *Grand Strategy*, volume 1, *Rearmament*. History of the Second World War. London: HMSO, 1976.

Giddens, Anthony. *The Consequences of Modernity*. Cambridge: Polity, 1990.

Gilbert, Martin. *Winston S. Churchill, 1922–1939: Prophet of Truth*. London: William Heinemann, 1976.

Gladstone, David. *The Twentieth-Century Welfare State*. Basingstoke: Macmillan, 1999.

Goode, G. Brown. 'The Principles of Museum Administration'. In *Museum Provision and Professionalism*, edited by Gaynor Kavanagh. London: Routledge, 1994, 41–46.

Gorman, G. E. and Sydney J. Shep, editors. *Preservation Management for Libraries, Archives and Museums*. London: Facet, 2006.

Goss, Chris, with Peter Cornwell and Bernd Rauchbach. *Luftwaffe Fighter-Bombers Over Britain: The Tip and Run Campaign, 1942–1943*. Mechanicsburg, PA: Stackpole, 2003.

Goss, Tracy; Richard Pascale and Anthony Athos. 'The Reinvention Roller Coaster: Risking the Present for a Powerful Future'. *Harvard Business Review* 71 (1993): 97–108.

Gottlieb, Julie. 'Umbrellas, Hats, and Gas Masks: The Objects of the War of Nerves, Britain 1938–1939'. Keynote address, Objects In and After Hostility: The Materiality of Conflict, Newcastle upon Tyne, 30–31 March 2017.

Graf, Rüdiger and Konrad H. Jarausch. '"Crisis" in Contemporary History and Historiography'. *Docupedia-Zeitgeschichte* (2017): https://doi.org/10.14765/zzf.dok.2.789.v1, https://docupedia.de/zg/Graf_jarausch_crisis_en_2017.

Gramsci, Antonio. *Selection from the Prison Notebooks*, edited and translated by Quintin Hoare and Geoffrey Nowell Smith. London: Lawrence and Wishart, 1971.

Gray, Clive. 'Instrumental Policies: Causes, Consequences, Museums and Galleries'. *Cultural Trends* 17 (2008): 209–222.

Gray, Clive. *The Politics of Museums*. Basingstoke: Palgrave Macmillan, 2015.

Green, Leanne. 'Advertising War: Pictorial Publicity, 1914–1918'. Doctoral thesis, Manchester Metropolitan University, 2015.

Greenhalgh, Paul. 'Education, Entertainment and Politics: Lessons from the Great International Exhibitions'. In *The New Museology*, edited by Peter Vergo. London: Reaktion, 1989, 74–98.

Griseri, Paul. *An Introduction to the Philosophy of Management*. London: SAGE, 2013.

Gurian, Elaine Heumann. 'Preface'. In *Institutional Trauma: Major Change in Museums and Its Effect on Staff*, edited by Elaine Heumann Gurian. Washington, DC: American Association of Museums, 1995, 17–21.

Hale, Joanne E.; David P. Hale and Ronald E. Dulek. 'Decision Processes during Crisis Response: An Exploratory Investigation'. *Journal of Managerial Issues* 18 (2006): 301–320.

Hall, Stuart. 'The Work of Representation'. In *Representation*, edited by Stuart Hall, Jessica Evans and Sean Nixon. 2nd edition. London: SAGE, 2013, 1–47.

Hall, Thomas and Wally Bock. *Ruthless Focus: How to Use Key Core Strategies to Grow Your Business*. Indianapolis, IN: Dog Ear, 2010.

Hallett, Tim and Marc J. Ventresca. 'Inhabited Institutions: Social Interactions and Organizational Forms in Gouldner's "Patterns of Industrial Bureaucracy"'. *Theory and Society* 35 (2006): 213–236.

Han, Ziqiang and Qiujie Zhang. 'Disruption of Organizations'. In *Encyclopedia of Crisis Management*, edited by K. Bradley Penuel, Matt Statler and Ryan Hagen. Thousand Oaks, CA: SAGE, 2013, 292–294.

Hannagan, Tim. *Mastering Strategic Management*. Basingstoke: Palgrave Macmillan, 2002.

Hannan, Leonie and Sarah Longair. *History Through Material Culture*. Manchester: Manchester University Press, 2017.

Hatim, Basil and Jeremy Munday. *Translation: An Advanced Resource Book*. Abingdon: Routledge, 2004.

Hatton, Alf. 'Current Issues in Museum Training In the UK'. In *Museum Management*, edited by Kevin Moore. London: Routledge, 1994, 148–155.

Hayes, John. *The Theory and Practice of Change Management*. 5th edition. London: Palgrave Macmillan, 2018.

Heal, Sharon. 'Editorial'. *Museums Journal* 114, number 4 (2014): 4.

Hein, Hilde S. *The Museum in Transition: A Philosophical Perspective*. Washington, DC: Smithsonian Institution Press, 2000.

Hill, Kate. 'Why Study Museum History?' Blog. Museums and Galleries History Group. https://www.mghg.info/blog/2017/5/27/why-study-museum-history.

Hill, Laura. 'Sadness as Durham Light Infantry Museum Closes Its Doors for the Final Time'. *Chronicle*, 31 March 2016. https://www.chroniclelive.co.uk/news/north-east-news/gallery/sadness-durham-light-infantry-museum-11121791.

Hillson, Franklin J. 'Barrage Balloons for Low Level Air Defence'. *Airpower Journal* 3 (1989): 27–40.

Hobbes, Thomas. *Leviathan; or, The Matter, Forme and Power of a Common Wealth Ecclesiasticall and Civil.* London: Andrew Crooke, 1651.

Hoffman, Andrew J. *From Heresy to Dogma: An Institutional History of Corporate Environmentalism.* Expanded edition. Stanford, CA: Stanford Business, 2001.

Holman, Brett. *The Next War in the Air: Britain's Fear of the Bomber, 1908–1941.* Abingdon: Routledge, 2016.

Hooper-Greenhill, Eilean. *Museums and the Shaping of Knowledge.* London: Routledge, 1992.

Hooper-Greenhill, Eilean. Museums and their Visitors. London: Routledge, 1994.

Hopkin, Paul. *Fundamentals of Risk Management: Understanding, Evaluating and Implementing Effective Risk Management.* 4th edition. London: Kogan Page, 2017.

Hopkins, Andrew. *Learning from High Reliability Organisations.* Sydney, NSW: CCH Australia, 2009.

Hornblower, Simon; Antony Spawforth and Esther Eidinow, editors. *The Oxford Classical Dictionary.* 4th edition. Oxford: Oxford University Press, 2012.

Horne III, John F.; and John E. Orr. 'Assessing Behaviours That Create Resilient Organizations'. *Employment Relations Today* 24 (1998): 29–39.

Hudson, Kenneth (1998). 'The Museum Refuses to Stand Still'. *Museum International* 50 (1): 43–50.

Hulton, John. *The Gunners of 1914: Baptism of Fire.* Barnsley: Pen and Sword Military, 2014.

Imperial War Museum. *A Short Guide to the Imperial War Museum.* 3rd edition. [London?]: [HMSO?], [1946?].

'Imperial War Museum'. *Manchester Guardian*, 10 June 1920.

International Council of Museums. *Code of Ethics for Museums.* Paris: International Council of Museums [2004?]. https://icom.museum/wp-content/uploads/2018/07/ICOM-code-En-web.pdf.

IWM. 'Annual Report and Accounts, 2020–2021'. [London?]: HMSO, 2021. https://assets.publishing.service.gov.uk/media/61bcbf50d3bf7f054e76962c/CCS0921374640-001__Imperial_War_Museum_Web_Accessible__5_.pdf.

IWM. 'Annual Report and Accounts, 2022–2023'. [London?]: HMSO, 2023. https://assets.publishing.service.gov.uk/government/uploads/system/uploads/attachment_data/file/1178402/Imperial_War_Museum_Annual_Report_and_Accounts_2022_to_2023.pdf.

IWM. 'Churchill War Rooms'. https://www.iwm.org.uk/visits/churchill-war-rooms.

IWM. 'HMS Belfast'. https://www.iwm.org.uk/visits/hms-belfast.

IWM. 'IWM Duxford'. https://www.iwm.org.uk/visits/iwm-duxford.

IWM. 'IWM Launches New Brand'. https://www.iwm.org.uk/sites/default/files/press-release/New_Brand_Press_Release.pdf.

IWM. 'IWM London'. https://www.iwm.org.uk/visits/iwm-london.

Jaeger, Stephen. *The Second World War in the Twenty-First-Century Museum: From Narrative, Memory and Experience to Experientiality.* Berlin, BE: Walter de Gruyter, 2020.

Janes, Robert R. 'Museums and the End of Materialism'. In *The Routledge Companion to Museum Ethics: Redefining Ethics for the Twenty-First Century Museum*, edited by Janet Marstine. Abingdon: Routledge, 2011, 54–69.

Janes, Robert R. *Museums and the Paradox of Change: A Case Study in Urgent Adaption.* 3rd edition. Abingdon: Routledge, 2013.

Janes, Robert R. *Museums in a Troubled World: Renewal, Irrelevance or Collapse?* Abingdon: Routledge, 2009.

Janes, Robert R. *Museums Without Borders: Selected Writings of Robert R. Janes.* Abingdon: Routledge, 2016.

Johnson, Allan G. *The Blackwell Dictionary of Sociology: A User's Guide to Sociological Language.* 2nd edition. Maldon, MA: Blackwell, 2000.

Johnson, Elliott; David Walker and Daniel Gray. *Historical Dictionary of Marxism.* 2nd edition. Lanham, MD: Rowman and Littlefield, 2014.

Johnston, Douglas A. 'Directors and Trustees: Personal Liability for Their Actions'. *The International Journal of Museum Management and Curatorship* 7 (1988): 77–78.

Jones, Gareth R. *Organizational Theory: Text and Cases.* 3rd illustrated edition. Upper Saddle River, NJ: Pretence-Hall, 2001.

Jupille, Joseph Henri and James A. Caporaso. *Theories of Institutions.* Cambridge: Cambridge University Press, 2022.

Kahneman, Daniel and Gary Klein. 'Conditions for Intuitive Expertise: A Failure to Disagree'. *American Psychologist* 64 (2009): 515–526.

Kahneman, Daniel. *Thinking, Fast and Slow.* London: Penguin, 2011.

Kaminsky, Graciela; Saul Lizondo; and Carmen Reinhart. 'Leading Indicators of Currency Crisis'. *IMF Staff Papers* 45 (1998): 1–48.

Kanter, Rosabeth Moss; Barry A. Stein and Todd D. Jick. *The Challenge of Organisational Change.* New York, NY: Free, 1992.

Kaplan, Flora E. S. 'Exhibitions as Communicative Media'. In *Museum, Media, Message*, edited by Eilean Hooper-Greenhill. Paperback edition. Abingdon: Routledge, 1999, 37–58.

Kavanagh, Gaynor. 'History in Museums in Britain: A Brief Survey of Trends and Ideas'. In *Social History in Museums: A Handbook for Professionals*, edited by David Fleming, Crispin Paine and John G. Rhodes. London: HMSO, 1993, 13–24.

Kavanagh, Gaynor. 'Museum as Memorial: The Origins of the Imperial War Museum'. *Journal of Contemporary History* 23 (1988): 77–97.

Kavanagh, Gaynor. *Museums and the First World War: A Social History.* London: Leicester University Press, 1994.

Keene, Suzanne. *Fragments of the World: Uses of Museum Collections.* Oxford: Elsevier Butterworth-Heinemann, 2005.

Kelly, Lynda. 'Measuring the Impact of Museums on Their Communities: The Role of the 21st Century Museum'. In *New Roles and Missions of Museums: INTERCOM 2006 Symposium*, edited by Ying-Ying Lai. Taipei: ICOM-INTERCOM, 2006, 25–34.

Kendra, James M. and Tricia Wachtendorf. 'Elements of Resilience After the World Trade Center Disaster: Reconstituting New York City's Emergency Operations Centre'. *Disasters* 27 (2003): 37–53.

Kenyon, J. 'Collecting for the Twenty-First Century'. In *Collections Management*, edited by Anne Fahy. London: Routledge, 1994, 122–125.

Khalil, Elias L. 'Organizations Verses Institutions'. *Journal of Institutional and Theoretical Economics* 151 (1995): 445–466.

'King and Queen at the War Museum'. *Daily Telegraph*, 10 June 1920.

Klein, Gary. *Sources of Power: How People Make Decisions.* Paperback edition. Cambridge, MA: MIT Press, 1999.

Klein, Richard J. T.; Robert J. Nicholls and Frank Thomalla. 'Resilience to Natural Hazards: How Useful Is This Concept?' *Environmental Hazards* 5 (2003): 33–45.

Knell, Simon J.; Suzanne Macleod and Sheila Watson. 'Introduction'. In *Museum Revolutions: How Museums Change and Are Changed*, edited by Simon J. Knell, Suzanne Macleod and Sheila Watson. Abingdon: Routledge, 2007, xix–xxvi.

Knight, Alan. '"Crisis" and the Great Depression in Latin America'. In *Crises in Economic and Social History: A Comparative Perspective*, edited by A. T. Brown, Andy Burn and Rob Doherty. Woodbridge: Boydell, 2015, 27–52.

Kolb, David A. *Experiential Learning: Experience as the Source of Learning and Development.* Upper Saddler River, NJ: Pearson Education, 2014.

Kölbel, Max. 'Faultless Disagreement'. *Proceedings of the Aristotelian Society* 104 (2004): 53–73.

Komatsuka, Carol M. 'Expanding the Museum audience Through Visitor Research'. In *Museum Management and Marketing*, edited by Richard Sandell and Robert R. Janes. London: Routledge, 2007, 366–373.

Koselleck, Reinhart. 'Crisis', translated by Michaela W. Richter. *Journal of the History of Ideas* 67 (2006): 357–400.

Koster, Emlyn H. and Stephen H. Baumann. 'Liberty Science Center in the United States: A Mission Focussed on External Relevance'. In *Looking Reality in the Eye: Museums and Social Responsibility*, edited by Robert R. Janes and Gerald T. Conty. Calgary, AB: University of Calgary Press, 2005, 85–111.

Kotter, John P. 'Leading Change: Why Transformation Efforts Fail'. In *Museum Management and Marketing*, edited by Richard Sandell and Robert R. Janes. Abingdon: Routledge, 2007, 20–29.

Lagadec, Patrick. *Preventing Chaos in a Crisis: Strategies for Prevention, Control and Damage Limitation*, translated by Jocelyn M. Phelps. London: McGraw-Hill, 1993.

Lahlou, Saadi. *Installation Theory: The Societal Construction and Regulation of Behaviour*. Cambridge: Cambridge University Press, 2018.

Landsborough, Gordon. *The Battle of the River Plate: The First Naval Battle of the Second World War*. Barnsley: Frontline, 2016.

Lang, Anthea. 'Shipley Bequest: Joseph Shipley and His Art Gallery'. [Gateshead?]: privately published, n.d.

Langmead, Louise and Peter Irving. *Inflammatory Bowel Disease*. Oxford: Oxford University Press, 2008.

Latchford, Peter. *The Future of Civic Museums: A Think Piece*. No place: Black Radley, 2018). https://www.nationalmuseums.org.uk/media/documents/publications/civic_museums_think_piece.pdf.

Law, Cheryl. *Suffrage and Power: The Women's Movement, 1918–1928*. London: I. B. Tauris, 1997.

Law, Jonathan. *A Dictionary of Business and Management*. 6th edition. Oxford: Oxford University Press, 2016.

Leege, David C. 'Catholics and the Civic Order: Parish Participation, Politics, and Civic Participation' *The Review of Politics* 50 (1988): 704–736.

Leflar, James J. and Marc H. Siegel. *Organizational Resilience: Managing the Risks of Disruptive Events – A Practitioner's Guide*. Boca Raton, FL: CRC, 2013.

Lengnick-Hall, Cynthia A. and Tammy E. Beck, 'Resilience Capacity and Strategic Agility: Prerequisites for Thriving in a Dynamic Environment'. In *Resilience and Engineering*. Resilience Engineering Perspectives 2, edited by Christpher P. Nemeth, Erik Hollnagel and Sidney Dekker. Farnham: Ashgate, 2009, 39–70.

Lengnick-Hall, Cynthia A.; Tammy E. Beck and Mark L. Lengnick-Hall. 'Developing A Capacity for Organizational Resilience Through Strategic Human Resource Management'. *Human Resource Management Review* 21 (2011): 243–255.

Lerbinger, Otto. *The Crisis Manager: Facing Disasters, Conflicts, and Failures*. 2nd edition. New York, NY: Routledge, 2012.

Lévi-Strauss, Claude. *The Savage Mind*. Chicago, IL: University of Chicago Press, 1966.

Lewis, Geoffrey. *For Instruction and Recreation: A Centenary History of the Museums Association*. London: Quiller, 1989.

Linkner, Josh. *The Road to Reinvention: How to Drive Disruption and Accelerate Transformation*. San Francisco, CA: Jossey-Bass, 2014.

Loewenstein, George. 'The Psychology of Curiosity: A Review and Reinterpretation'. *Psychological Bulletin* 116 (1994): 75–98.

Longmate, Norman. *How We Lived Then: A History of Every Day Life During the Second World War*. London: Pimlico, 2002.

Lord, Gail Dexter. 'The Exhibition Planning Process'. In *The Manual of Museum Exhibitions*, edited by Barry Lord and Gail Dexter Lord. Walnut Creek, CA: AltaMira, 2001, 1–8.

Lowenthal, David. 'Patrons, Populists, Apologists: Crises in Museum Stewardship'. In *Valuing Historic Environments*, edited by Lisanne Gibson and John Pendlebury. Farnham: Ashgate, 2009, 19–31.

Loyear, Rachelle. *The Manager's Guide to Simple, Strategic, Service-Oriented Business Continuity*, edited by Kristen Noakes-Fry. No place: Rothstein, 2017. https://www.rothstein.com/product/simple-strategic-service-oriented-business-continuity-management/.

Luckham, Robin. 'Armament Culture'. *Alternatives* 10 (1984): 1–44.

Mackay, Robert. *Half the Battle: Civilian Morale in Britain during the Second World War*. Manchester: Manchester University Press, 2002.

Maclean, Mairi; Charles Harvey and Stewart R. Clegg. 'Conceptualizing Historical Organizational Studies'. *Academy of Management Review* 41 (2016): 609–632.

Madison, H. L. 'Tentative Code of Museum Ethics'. In *Museum Provision and Professionalism*, edited by Gaynor Kavanagh. London: Routledge, 1994, 265–270.

Madison, James H. *Slinging Doughnuts for the Boys: An American Woman in World War II*. Bloomington, IN: Indiana University Press, 2007.

Maguire, Brigit and Sophie Cartwright. *Assessing a Community's Capacity to Manage Change: A Resilience Approach to Social Assessment*. Canberra, ACT: Bureau of Rural Science, 2008. https://www.agriculture.gov.au/sites/default/files/abares/documents/dewha-resilience-sa-report-final-4.pdf.

Mallak, Larry A. 'Measuring Resilience in Health Care Provider Organizations'. *Health Manpower Management* 24 (1998): 148–152.

Malvern, Sue. 'War, Memory and Museums: Art and Artefact in the Imperial War Museum'. *History Workshop Journal* 49 (2000): 177–203.

Malvern, Sue. *Modern Art, Britain and the Great War: Witnessing, Testimony and Remembrance*. New Haven, CT: Yale University Press, 2004.

March, James; Erhard Friedberg and David Arellano. 'Institutions and Organizations: Differences and Linkages from Organization Theory'. *Gestión y Política Pública* 20 (2011): 235–246.

Marcovitch, Harvey. *Black's Medical Dictionary*. 43rd edition. London: Bloomsbury Information, 2017.

Markham, S. F. *A Report on the Museums and Art Galleries of the British Isles (Other Than the National Museums)*. Edinburgh: Carnegie United Kingdom Trustees, 1938.

Martin, Dan. 'Please? Please? Please? Meh: Beatles Museum Closes Owing to Lack of Interest'. *Guardian*, 7 June 2012. https://www.theguardian.com/music/2012/jun/07/beatles-museum-hamburg.

Martin, Ron and Peter Sunley. 'On the Notion of Regional Economic Resilience: Conceptualization and Explanation'. *Journal of Economic Geography* 15 (2015): 1–42.

Marwick, Arthur, editor. *Total War and Social Change*. Basingstoke: Macmillan, 1988.

Marwick, Arthur. *War and Social Change in the Twentieth Century*. London: Macmillan, 1974.

Marwick, Arthur; Clive Emsley and Wendy Simpson, editors. *Total War and Historical Change: Europe, 1914–1955*. Maidenhead: Open University Press, 2001.

Mason, Rhiannon, editor. *Museum Studies*. 5 volumes. Abingdon: Routledge, 2019.

Mason, Rhiannon. 'Museums, Galleries and Heritage: Sites of Meaning-making and Communication'. In *Heritage, Museums and Galleries: An Introductory Reader*, edited by Gerard Corsane. Abingdon: Routledge, 2005, 200–214.

Mason, Rhiannon; Alistair Robinson and Emma Coffield. *Museum and Gallery Studies: The Basics*. Abingdon: Routledge, 2018.

Massey, Joseph Eric. 'Managing Organizational Legitimacy: Communication Strategies for Organizations in Crisis'. *The Journal of Business Communication* 38 (2001): 153–183.

Matassa, Freda. *Valuing Your Collection: A Practical Guide for Museums, Libraries and Archives*. London: Facet, 2017.

Mathenge, Joseph. 'Business Continuity vs Business Resiliency: What's the Difference?' The Business of IT Blog. BMC. https://www.bmc.com/blogs/business-continuity-vs-resiliency/.

Matthews, Graham; Yvonne Smith and Gemma Knowles. *Disaster Management in Archives, Libraries and Museums*. Aldershot: Ashgate, 2009.

McCamley, N. J. *Saving Britain's Art Treasures*. Barnsley: Leo Cooper, 2003.

McCarthy, Ian P.; Mark Collard and Michael Johnson. 'Adaptive Organizational Resilience: An Evolutionary Perspective'. *Current Opinion in Environmental Sustainability* 28 (2017): 33–40.

McClintock, Anne. *Imperial Leather: Race, Gender and Sexuality in the Colonial Contest*. New York, NY: Routledge, 1995.

McLean, Fiona. *Marketing the Museum*. London: Routledge, 1997.

McManus, Sonia; Erica Seville; John Vargo and David Brunsdon. 'Facilitated Process for Improving Organizational Resilience'. *Natural Hazards Review* 9 (2008): 81–90.

Mercer, Alyson. 'The Changing Face of Exhibiting Women's Wartime Work at the Imperial War Museum'. *Women's History Review* 22 (2013): 330–344.
Meyer, John W. and Brian Rowan. 'Institutionalized Organizations: Formal Structure as Myth and Ceremony'. *American Journal of Sociology* 83 (1977): 340–363.
Miers, Henry. *A Report on the Public Museums of the British Isles (Other Than the National Museums)*. Edinburgh: Carnegie United Kingdom Trustees, 1928.
Miller, Steven. *The Anatomy of a Museum: An Insider's Text*. Hoboken, NJ: Wiley Blackwell, 2018.
Miller-Cole, Bianca and Byron Cole. *Self Made: The Definitive Guide to Business Startup Success*. London: John Murray Learning, 2017.
Milosch, Jane C. and Nick Pearce, editors. *Collecting and Provenance: A Multidisciplinary Approach*. Lanham, MD: Rowman and Littlefield, 2019.
Milstein, Brian. 'Thinking Politically About Crisis: A Pragmatist Perspective'. *European Journal of Political Theory* 14 (2015): 141–160.
Monger, David. *Patriotism and Propaganda in the First World War: The National War Aims Committee and Civilian Morale*. Liverpool: Liverpool University Press, 2012.
Monti, Francesca and Suzanne Keene. *Museums and Silent Objects: Designing Effective Exhibitions*. Abingdon: Routledge, 2016.
Morris, Sammie L. 'Starting from Scratch: How to Create a Museum Archives'. *Libraries Research Publications* 45: https://docs.lib.purdue.edu/cgi/viewcontent.cgi?article=1052&context=lib_research.
Moss, Stephen. 'The Imperial War Museum: As Much a Relic as Its Spitfires and Doodlebugs?' *Guardian*, 10 July 2014. https://www.theguardian.com/culture/2014/jul/10/imperial-war-museum-reopens-first-world-war.
Mouzakitis, Angelos. 'Modernity and the Idea of Progress'. *Frontiers in Sociology* 2 (2017): https://doi.org/10.3389/fsoc.2017.00003, https://www.frontiersin.org/articles/10.3389/fsoc.2017.00003/full.
'Mr L. R. Bradley'. *Times*, 30 January 1968.
'Mr. Churchill in Manchester'. *Manchester Guardian*, 29 January 1940.
Ndlela, Martin N. *Crisis Communication: A Stakeholder Approach*. Cham: Springer Nature, 2019.
Nel, Philip and Marjolein Righarts. 'Natural Disasters and the Risk of Violent Civil Conflict'. *International Studies Quarterly* 52 (2008): 159–185.
Nesbit, Rebecca and David A. Reingold. 'Soldiers to Citizens: The Link between Military Service and Volunteering'. *Public Administration Review* 71 (2011): 67–76.
'News from the Museums'. *Museums Journal* 36 (1936): 263–265.
'News from the Museums'. *Museums Journal* 37 (1938): 521–526.
'News from the Museums'. *Museums Journal* 38 (1938): 199–204.

Nixon, Barbra. *Raiders Overhead: A Diary of the London Blitz*, edited by Ian Grant. London: Scolar, 1980.
North, Douglas. *Institutions, Institutional Change and Economic Performance*. Cambridge: Cambridge University Press, 1990.
O'Brien, Terence H. *Civil Defence*. History of the Second World War. London: HMSO, 1955.
O'Keefe, Roger. *The Protection of Cultural Property in Armed Conflict*. Cambridge: Cambridge University Press, 2006.
Olusoga, David. *Black and British: A Forgotten History*. Paperback edition. London: Pan, 2017.
Orlikowski, Wanda J. 'Material Knowing: The Scaffolding of Human Knowledgeability'. *European Journal of Information Systems* 15 (2006): 460–466.
Overy, R. J. *The Inter-war Crisis, 1919–1939*. Harlow: Pearson, 1994.
Overy, Richard. *The Bombing War: Europe, 1939–1945*. Paperback edition. London: Penguin, 2014.
Paddon, Hannah. *Redisplaying Museum Collections: Contemporary Display and Interpretation in British Museums*. Farnham: Ashgate, 2014.
Paine, Thomas. *The American Crisis: Number 1*. [Philadelphia, PA?]: [Styner and Cist?], 1776.
Panter-Downes, Mollie. *London War Notes, 1939–1945*, edited by William Shawn. London: Longman, 1972.
Paraskevas, Alexandros. 'Mitroff's Five Stages of Crisis Management'. In *Encyclopedia of Crisis Management*, edited by K. Bradley Penuel, Matt Statler and Ryan Hagen. Thousand Oaks, CA: SAGE, 2013, 628–632.
Paraskevas, Alexandros. 'Risk Treatment'. In *Encyclopedia of Crisis Management*, edited by K. Bradley Penuel, Matt Statler and Ryan Hagen. Thousand Oaks, CA: SAGE, 2013, 842–845.
Parsons III, Thad. 'The Science Museum and the Second World War'. In *Science for the Nation: Perspectives on the History of the Science Museum*, edited by Peter J. T. Morris. Basingstoke: Palgrave Macmillan, 2010, 61–89.
Pathak, Harsh. *Organisational Change*. New Delhi, NCT: Pearson Education, 2011.
'"Peace in Our Time" in the War Museum'. *Daily Mail*, 7 February 1940.
Pearce, Susan M. *Museums, Objects and Collections: A Cultural History*. London: Leicester University Press, 1992.
Pearson, Catherine. *Museums in the Second World War: Curators, Culture and Change*, edited by Suzanne Keene. Abingdon: Routledge, 2017.
Pearson, Christine M. and Ian I. Mitroff. 'From Crisis Prone to Crisis Prepared: A Framework for Crisis Management'. *The Executive* 7 (1993): 48–59.
Perry, James L. and Ann Marie Thomson. *Civic Service: What Difference Does It Make?* Abingdon: Routledge, 2015.

Pietersen, Willie. *Reinventing Strategy: Using Strategic Learning to Create and Sustain Breakthrough Performance*. New York, NY: John Wiley.

Plaza, Beatriz and Silke N. Haarich. 'Museums for Urban Regeneration? Exploring Conditions for Their Effectiveness'. *Journal of Urban Regeneration and Renewal* 2 (2009): 259–271.

Powell, Brent A. *Collection Care: An Illustrated Handbook for the Care and Handling of Cultural Objects*. Lanham, MD: Rowman and Littlefield, 2016.

Prior, Nick. 'Postmodern Restructurings'. In *A Companion to Museum Studies*, edited by Sharon Macdonald. Maldon, MA: Blackwell, 2006, 509–524.

Prior, Robin and Trevor Wilson. *The Somme*. New Haven, CT: Yale University Press, 2005.

Pursiainen, Christer. *The Crisis Management Cycle*. Abingdon: Routledge, 2018.

Quarantelli, E. L. 'Disaster Research Center'. In *Encyclopedia of Sociology*, volume 1, edited by Edgar F. Borgatta and Rhonda J. V. Montgomery. 2nd edition. New York: Macmillan Reference USA, 2000, 681–688.

Raco, Mike and Emma Street. 'Resilience Planning, Economic Change and The Politics of Post-recession Development in London and Hong Kong'. *Urban Studies* 49 (2011): 1065–1087.

Ravail, Florian. 'The Crisis: Meaningless Catch-all Term or Rigorous Concept?', translated by Thomas Williams. *Kinea* (2018). Note that this reference is no longer available as the online publication has been discontinued.

Ray, John. *The Night Blitz, 1940–1941*. New edition. London: Cassell Military, 2000.

Redman, Samuel J. *The Museum: A Short History of Crisis and Resilience*. New York, NY: New York University Press, 2022.

Rees, Nigel. *Don't You Know There's a War On? Words and Phrases from the World Wars*. London: Batsford, 2011.

Reeve, John. 'Prioritizing Audience Groups'. In *The Responsive Museum: Working with Audiences in the Twenty-First Century*, edited by Caroline Lang, John Reeve and Vicky Woolard. Aldershot: Ashgate, 2006, 43–60.

Reynolds, David. *The Long Shadow: The Great War and the Twentieth Century*. Paperback edition. London: Simon and Schuster, 2014.

Rhys, Owain and Zelda Baveystock. 'Introduction'. In *Collecting the Contemporary: A Handbook for Social History Museums*, edited by Owain Rhys and Zelda Baveystock. Edinburgh: MuseumsEtc, 2014, 15–36.

Rhys, Owain. *Contemporary Collecting: Theory and Practice*. Edinburgh: MuseumsEtc, 2011.

Richardson, Bill. 'Socio-technical Disasters: Profile and Prevalence'. *Disaster Prevention and Management* 3 (1994): 41–69.

Robinson, John Martin. *Requisitioned: The British Country House in the Second World War*. London: Arum, 2014.

Rochlin, Gene I. 'Informal Organisational Networking as a Crisis Avoidance Strategy: US Naval Flight Operations as a Case Study'. *Industrial Crisis Quarterly* 3 (1989): 159–176.

Rochlin, Gene I., editor. *Journal of Contingencies and Crisis Management* 4 (1996): 55–112.

Roe, Emily and Paul R. Schulman. *High Reliability Management: Operating on the Edge*. Stanford, CA: Stanford University Press, 2008.

Romero Salvadó, Francisco J. *Historical Dictionary of the Spanish Civil War*. Lanham, MD: Scarecrow, 2013.

Royal Commission on National Museums and Galleries. *Final Report, Part I: General Conclusions and Recommendations*. London: HMSO, 1929.

Royal Commission on National Museums and Galleries. *Final Report, Part II: Conclusions and Recommendations Relating To Individual Institutions*. London: HMSO, 1930.

Royal Commission on National Museums and Galleries. *Interim Report*. London: HMSO, 1928.

Saleh, Yunus D. *Crisis Management: The Art of Success and Failure—30 Case Studies in Business and Politics*. Minneapolis, MN: Mill City, 2016.

Schlatter, N. Elizabeth. *Museum Careers: A Practical Guide for Students and Novices*. Abingdon: Routledge, 2016.

Scott, Carol A., editor. *Museums and Public Value: Creating Sustainable Futures*. Abingdon: Routledge, 2016.

Scott, James. 'Objects and the Representation of War in Military Museums'. *Museum and Society* 13 (2015): 489–502.

Scott, W. Richard. *Institutions and Organizations: Ideas, Interests and Identities*. 4th edition. Thousand Oakes, CA: SAGE, 2014.

Scott, W. Richard; John W. Meyer and associates. *Institutional Environments and Organizations: Structural Complexity and Individualism*. Thousand Oakes, CA: SAGE, 1994.

Selznick, Philip. *Leadership in Administration: A Sociological Interpretation*. Berkeley, CA: University of California Press, 1984.

Seville, Erica. *Resilient Organizations: How to Survive, Thrive and Create Opportunities through Crisis and Change*. London: Kogan Page, 2017.

Sewell Jr, William H. 'A Theory of Structure: Duality, Agency, and Transformation'. *American Journal of Sociology* 98 (1992): 1–29.

Shaw, H. 'Science Museums'. *Museums Journal* 46 (1946): 173.

Shaw, Keith and Louise Maythorne. 'Managing for Local Resilience: Towards a Strategic Approach'. *Public Policy and Administration* 28 (2011): 43–65.

Shenton, Caroline. *National Treasures: Saving the Nation's Art in World War II*. Paperback edition. London: John Murray, 2022.

Sheppard, Stephen. 'Museums in the Neighborhood: The Local Economic Impact of Museums'. In *Handbook of Industry Studies and Economic Geography*, edited by Frank Giarratani, Geoffrey J. D. Hewings and Philip McCann. Cheltenham: Edward Elgar, 2013, 191–204.

Sherraden, Michael Wayne and Donald J. Eberly. 'Why National Service?' In *National Service: Social, Economic, and Military Impacts*, edited by Michael Wayne Sherraden and Donald J. Eberly. Elmsford, NY: Pergamon, 1982, 3–20.

Silvia, Paul J. 'Curiosity and Motivation'. In *The Oxford Handbook of Human Motivation*, edited by Richard M. Ryan. Oxford: Oxford University Press, 2012, 157–166.

Simkins, Peter. 'The Four Armies, 1914–1918'. In *The Oxford History of the British Army*, edited by David G. Chandler with the assistance of Ian Beckett. Oxford: Oxford University Press, 1996, 235–255.

Simmons, John E. *The Museum: A History*. Lanham, MD: Rowman and Littlefield, 2016.

Singleton, Rosalind Watkiss. "Doing Your Bit': Women and the National Savings Movement in the Second World War'. In *The Home Front in Britain: Images, Myths and Forgotten Experiences since 1914*, edited by Maggie Andrews and Janis Lomas. Basingstoke: Palgrave Macmillan, 2014, 217–231.

'Sir Bertram Cubitt'. *Times*, 24 September 1942.

Smither, Roger and David Walsh. 'Unknown Pioneer: Edward Foxen Cooper and the Imperial War Museum Film Archive, 1919–1934'. *Film History* 12 (2000): 187–203.

Snow, David A.; E. Burke Rochford Jr; Steven K. Worden and Robert D. Benford. 'Frame Alignment Processes, Micromobilization, and Movement Participation'. *American Sociological Review* 51 (1986): 464–481.

Sornette, Didier. 'Dragon-Kings, Black Swans and the Prediction of Crises'. International Journal of Terraspace Science and Engineering 2 (2009): https://doi.org/10.48550/arXiv.0907.4290, https://arxiv.org/ftp/arxiv/papers/0907/0907.4290.pdf.

Stake, Robert E. *The Art of Case Study Research*. Thousand Oaks, CA: SAGE, 1995.

Stammers, Neil. *Civil Liberties in Britain During the 2nd World War: A Political Study*. Abingdon: Routledge, 2023.

Standing Commission on Museums and Galleries. *Third Report: The War Years and After*. London: HMSO, 1948.

Stanford, Michael. *A Companion to the Study of History*. Oxford: Blackwell, 1994.

Stead, Eleanor and Clive Smallman. 'Understanding Business Failure: Learning and Un-learning Lessons from Industrial Crises'. *Understanding Business Failure* 7 (1999): 1–18.

Stewart, Dave. 'Founders Hall Interpretive Centre Closes Doors'. *SaltWire*, 10 May 2016. https://www.saltwire.com/prince-edward-island/news/founders-hall-interpretive-centre-closes-doors-101756/.

Strebel, Paul. 'Why Do Employees Resist Change?' *Harvard Business Review* 74 (1996): 86–92.

Suchman, Mark C. 'Managing Legitimacy: Strategic and Institutional Approaches'. *Academy of Management Review* 20 (1995): 571–610.

Suddaby, Roy; Alex Bitektine and Patrick Haack. 'Legitimacy'. *Academy of Management Annals* 11 (2017): 451–478.

Sukel, William M. 'Museums as Organisations'. In *Museum Management*, edited by Kevin Moore. London: Routledge, 1994, 262–264.

Sullivan, Dan. 'The Power of Entrepreneurship During A Crisis'. Multiplier Mindset Blog. Strategic Coach. https://resources.strategiccoach.com/the-multiplier-mindset-blog/the-power-of-entrepreneurship-during-a-crisis.

Sullivan, Nikki. 'Contemporary Collecting in a History Museum'. MuseumNext. https://www.museumnext.com/article/museums-contemporary-collecting/.

Taleb, Nassim Nicholas. *The Black Swan: The Impact of the Highly Improbable*. London: Allen Lane, 2007.

Taylor, James. 'Interpreting the Second World War'. In *The Power of Objects: Museums and World War II*, edited by Esben Kjeldbæk. Edinburgh: MuseumsEtc, 2009, 52–81.

Taylor, James. *'Your Country Needs You': The Secret History of the Propaganda Poster*. Glasgow: Saraband, 2013.

'The Greatest War Memorial'. *Times*, 10 June 1920.

'The Imperial War Museum Reopens'. *Illustrated London News*, 3 February 1940.

'The Imperial War Museum'. *Times*, 26 January 1940.

'The King Opens War Museum'. *Daily Mail*, 10 June 1920.

'The Museums Reopen'. *Museums Journal* 39 (1940): 454–455.

Thompson, Julian. *Dunkirk: Retreat to Victory*. London: Pan, 2009.

Thorsheim, Peter. *Waste into Weapons: Recycling in Britain during the Second World War*. New York, NY: Cambridge University Press, 2015.

Tomasiewicz, Kasia. '"We are a social history, not a military history museum": Large Objects and the "Peopling" of Galleries in the Imperial War Museum, London'. In *Museums, Modernity and Conflict: Museums and Collections in and of War Since the Nineteenth Century*, edited by Kate Hill. Abingdon: Routledge, 2021, 214–234.

Tonkin, Sam. 'Hitler's Plans to Destroy London: Rare Mare Map Revealing Germany's WWII Bombing Targets Is Discovered After 75 Years in An Attic'. *Mail Online*, 24 February 2017. https://www.dailymail.co.uk/news/article-4256244/Map-revealing-Germany-s-WWII-bombing-targets-discovered.html.

Tosh, John. *The Pursuit of History: Aims Methods and New Directions in the Study of Modern History*. 3rd edition. Harlow: Longman, 2002.

Tuck, Fiona and Scott Dickinson. *The Economic Impact of Museums in England*. Newcastle upon Tyne: TBR, 2015. https://www.artscouncil.org.uk/sites/default/files/download-file/Economic_Impact_of_Museums_in_England_report.pdf.

Tulliach, Anna. 'Best Practice in Museum Partnerships: Bologna and its Museum Networking'. *Museological Review* 21 (2017): 8–22.

Uhle-Wettler, Franz. 'War'. In *Brassey's Encyclopedia of Military History and Biography*, edited by Franklin D. Margiotta. Washington, DC: Brassey's, 1994, 1045–1064.

Van Wart, Montgomery. 'The First Step in the Reinvention Process: Assessment'. *Public Administration Review* 55 (1995): 429–438.

Veil, Shari R. 'Routine Emergencies versus True Crises'. In *Encyclopedia of Crisis Management*, edited by K. Bradley Penuel, Matt Statler and Ryan Hagen. Thousand Oaks, CA: SAGE, 2013, 845–847.

Vermeulen, Freek; Phanish Puranam and Ranjay Gulati. 'Change for Change's Sake'. *Harvard Business Review* 88 (2010): 70–76.

Vincent, Charles; Susan Burnett and Jane Carthey. *The Measurement and Monitoring of Safety: Drawing Together Academic Evidence and Practical Experience to Produce a Framework for Safety Measurement and Monitoring*. London: Health Foundation, 2013.

Vukajlović, Vladimir; Ivana Simeunović; Isidora Beraha; and Miodrag Brzaković. 'Importance of Information in Crisis Management – Statistical Analysis'. *Industrija* 47 (2019): 37–53.

Wallis, James and James Taylor. 'The Art of War Display: The Imperial War Museum's First World War Galleries, 2014'. In *Commemorative Spaces of the First World War: Historical Geographies at the Centenary*, edited by James Wallis and David C. Harvey. Abingdon: Routledge, 2018, 101–114.

Ward, Lawrence. *London County Council Bomb Damage Maps, 1939–1945*. London: Thames and Hudson, 2015.

Warner, Philip. *Firepower: From Slings to Star Wars*. London: Grafton, 1988.

Watson, Sheila. 'Museums and their Communities'. In *Museums and their Communities*, edited by Sheila Watson. Abingdon: Routledge, 2007, 1–23.

Watters, Jamie. *The Business Continuity Management Desk Reference: Guide to Business Continuity Planning, Crisis Management and IT Disaster Recovery*. No place: Business Leverage, 2010.

Weick, Karl E. 'The Collapse of Sensemaking in Organizations: The Mann Gulch Disaster'. *Administrative Science Quarterly* 38 (1993): 628–652.

Weick, Karl E. and Kathleen M. Sutcliffe. *Managing the Unexpected Resilient Performance in an Age of Uncertainty*. 3rd edition. Hoboken, NJ: John Wiley, 2015.

Weick, Karl E.; Kathhleen M. Sutcliffe and David Obstfeld. 'Organizing for High Reliability: Processes of Collective Mindfulness'. In *Research in Organizational Behaviour*, edited by Barry M. Staw and Robert I. Sutton, 21. Greenwich, CT: JAI, 1999, 81–123.

Weil, Stephen. *Making Museums Matter*. Washington, DC: Smithsonian Institution Press, 2002.

Weingärtner, Jörn. *The Arts as a Weapon of War: Britain and the Shaping of National Morale in World War II*. Revised paperback edition. London: I. B. Tauris, 2012.

Welch, David. 'Propaganda, Definitions of'. In *Propaganda and Mass Persuasion: A Historical Encycloedia, 1500 to the Present*, edited by Nicholas J. Cull, David Culbert and David Welch. Santa-Barbra, CA: ABC-CLIO, 2003, 317–323.

Wellington, Jennifer. *Exhibiting War: The Great War, Museums, and Memory in Britain, Canada, and Australia*. Cambridge: Cambridge University Press, 2017.

Wells, H. G. *The War That Will End War*. London: F. and C. Palmer, 1914.

Whitehead, Christopher; Mads Daugbjerg; Susannah Eckersley and Gönül Bozoğlu. 'Dimensions of European Heritage and Memory: A Framework Introduction'. In *Dimensions of Heritage and Memory: Multiple Europes and the Politics of Crisis*, edited by Christopher Whitehead, Susannah Eckersley, Mads Daugbjerg and Gönül Bozoğlu. Abingdon: Routledge, 2019, 1–25.

Whittaker, Debbie. 'Museums at War: The Impact of the Second World War on the National Museums of London, with Particular Reference to the Imperial War Museum, National Gallery, Victoria and Albert Museum and the British Museum'. Master's dissertation, University of Leicester, 2010.

Wilding, Norman and Philip Laundy. *An Encyclopaedia of Parliament*. 3rd edition. New York, N.Y.: Frederick A. Praeger, 1968.

Williamson, Oliver E. 'Transaction Cost Economics and Organization Theory'. *Industrial and Corporate Change* 2 (1993): 107–156.

Williamson, Oliver E. *Markets and Hierarchies—Analysis and Antitrust Implications: A Study in the Economics of Internal Organization*. New York, NY: Free, 1975.

Williamson, Oliver E. *The Economic Institutions of Capitalism: Firms, Markets, Relational Contracting*. New York, NY: Free, 1985.

Wilson, Deirdre and Dan Sperber. 'Relevance Theory'. In *The Handbook of Pragmatics*, edited by Laurence R. Horn and Gregory L. Ward. 2nd edition. Maldon, MA: Blackwell, 2006, 607–632.

Wilson, Jennifer. 'Now the Disaster's Happened, What Am I Supposed to Do?' Accounting Today. https://www.accountingtoday.com/news/now-the-disasters-happened-what-am-i-supposed-to-do.

Wucker, Michele. *The Gray Rhino: How to Recognize and Act on the Obvious Dangers We Ignore*. New York, NY: St Martin's, 2016.

Yin, Robert K. *Case Study Research: Design and Methods*. 5th edition. Thousand Oakes, CA: SAGE, 2014.

Zaloga, Steven J. *V-1 Flying Bomb, 1942–52*. Botley: Osprey, 2005.

Zaloga, Steven J. *V-2 Ballistic Missile, 1942–52*. Botley: Osprey, 2003.

Ziegler, Philip. *Between the Wars*. London: MacLehose, 2016.

Zimmerman, Monica A. and Gerald J. Zeitz. 'Beyond Survival: Achieving New Venture Growth by Building Legitimacy'. *Academy of Management Review* 27 (2002): 414–431.

Zucker, Lynne G. 'Organizations as Institutions'. In *Research in the Sociology of Organizations*, volume 2, edited by Samuel B. Bacharach. Greenwich, CT: JAI, 1983, 1–47.

Zweiniger-Bargielowska, Ina. *Austerity in Britain: Rationing, Controls, and Consumption, 1939–1955*. Oxford: Oxford University Press, 2000.

Index

P. W. Deans, *Crisis, Reinvention and Resilience in Museums*,
https://doi.org/10.1007/978-3-031-67806-6

GPSR Compliance

The European Union's (EU) General Product Safety Regulation (GPSR) is a set of rules that requires consumer products to be safe and our obligations to ensure this.

If you have any concerns about our products, you can contact us on ProductSafety@springernature.com

In case Publisher is established outside the EU, the EU authorized representative is:

Springer Nature Customer Service Center GmbH
Europaplatz 3
69115 Heidelberg, Germany

The manufacturer's authorised representative in the EU is Springer Nature Customer Service Centre GmbH, Europaplatz 3, 69115 Heidelberg, Germany. If you have any concerns regarding our products, please contact ProductSafety@springernature.com

Printed and bound by CPI Group (UK) Ltd, Croydon, CR0 4YY
28/09/2026
02224072-0012